I0394009

# A Tenure of Consciousness?

By

Damon Dion Reed

# Contents

# I. Squishy-Squish

The thumper inside my chest cavity goes bumpity-bump to the rhythm that my brain sets with all its squishing and squirting. But don't worry, this book isn't about that complex squishy organ inside your noggin, which facilitates your existence...hopefully. This book is about how the universe is made of positive and negative energy 'squishing' and 'squirting' on the atomic level. Unfortunately, before we get to all those pornographic terms, as it pertains to the perpetual existence of matter-battery-energy, we need to ponder and even more disturbing collection of words.

No matter how much we want to separate religion from science, these two unique entities will always be intertwined. For example, religious scientists believe that the universe will NOT die because God is eternal, which has resulted in the postulate that a NEW universe will rise from the ashes of the OLD universe. Supposedly, the OLD universe will stop expanding and collapse to create a second big bang that will create a NEW universe, as it is dictated by God's never ending love...or something like that. Unfortunately,

NON-religious scientists have invented lots of mumbo-jumbo words to describe gravity, entropy, and twerking-quirks in order to push God out of the universal theories...or something like that. Granted, I could be confused as to which scientific theories support God and/or the Devil, but I'm not really worried about all that hoopla. What I am worried about is logic. For starters, entropy would never exist if energy did NOT decay and expand, which is the reason why the universe is expanding. (If you keep reading, I'll explain **that** in more detail later.) Second, until recently, nobody has been able to give a reasonable explanation as to how gravity exists in our solar system. (Hint: It is all about the squishing and squirting of negative energy in our branch of the universe.) And finally, scientists believe that humans have the ENORMOUS potential to investigate EVERY inch of the universe. (That was a lie. NON-religious scientists think we're all illogical monkey descendants...or something like that?) In any event, let's assume that our ancestors were a little close minded, in more ways than one, and approach the metaphysical conundrum of our existence with a tad bit more logic. I'm not asking that we approach the grandiose-ness of this universe with ABSOLUTE logic, because that is HUMANLY **impossible**. All I'm asking is that we approach our existence with a smidge more logic...please?

I believe there is a God and that his existence is eternal. But having said that though, I do NOT think that God needs to collect the decaying remnants from an OLD universe in order to create a NEW universe. Personally, I think God prefers to have a clean canvas every time he/she/it creates a new universal work of art. Therefore, I religiously and scientifically **reject** that our universe will stop expanding, start contracting, and undergo a second big bang. Granted, I have severe reservations as is correlates to the "supposed" logic of the first big bang, but that is beside the point.

I believe the creation of this universe is based upon varying levels of highly order energy, which is SLOWLY degrading to create space/time and subsequently, entropy. Unfortunately, since I postulated that each part of the universe has variable concentrations of positive/negative energy and that the energy within each **branch** of the universe is not always stable in other branches of the universe, it is completely LOGICAL to postulate that we can NOT see the entire universe. All of which, will lead to an eternity of scientific and religious debate. (Thank Goodness! God has a long-term plan for humanity?)

Having said all that, I previously postulated that the universe consists of galaxies that behave like LARGE decaying battery systems, which complicates things when you associate that postulate with the postulate that there are varying levels of positivity/negativity in different branches of the universe. You see, even though there is a possible 'energy conservation' via the association of different universal branches that have varying **amounts** of positivity/negativity, which renders it impossible to observe adjacent universal branches, it still should be possible to observe energetic clouds that do not adhere to normal physics…as it pertains to the charge separation in our branch of the universe. All of which, is a kick to one's mental roll-cage when you think about it too much. In any event, let's return to the postulate that God likes a clean canvas when he/she/it creates something beautiful.

As **illogical** as it may seem that God can create a complex universal battery system that allows humans to exist on an infinitesimally small portion of the battery system, while being unable to see EVERY part of the very complex universal battery system, I choose to believe that God's plan results in the charge separation that creates the reoccurring laws that humans have incorporated into physics. Granted, we humans like to make massive assumptions, especially when it comes to gravity and the universe, but if our view of the universe is limited, then how can our beliefs-postulates-laws-universal-gravity-constants be universally correct?

In conclusion, I realize that it is a ridiculous idea to write a book about God's plan, especially if it pertains to science and/or our limited view of the universe, but I want a pinch more logic in my existence...even if it is nowhere close to God's logic. In any event, I enjoy musing about the intricacies of a Quanta Dynamic universe as it pertains to the forces of galactic coalescence caused by a universe of squishing, squirting, and decaying energyh. (All of which, I haven't even started talking about.) That is why I decided to write another book, which you are now reading. Unfortunately, since I haven't made enough money to sustain my existence with writing, this is a multiple book collection entitled: *A Tenure of Consciousness?* Hopefully, this shotgun approach to logic will earn me some money, which is the ultimate reason for writing...I think.

# Chapter 2

One fundamental scientific misunderstandings of past and present, beside the belief that God isn't able to make another universe from scratch, is the gravity-entropy conundrum. Granted, the conundrum of gravity-entropy is esthetically ENORMOUS when someone envisions matter as **never** degrading and gravity as a **magical** force that has nothing to do with the diffusion of positive or negative energy, but that is beside the point. The point is, when a logical base is provided for entropy-gravity, the conundrum enshrouding gravity-entropy dissipates like a mysterious magical medium monster, which is always enshrouding science and life. Therefore, let us begin with a slight review of entropy and gravity.

Entropy is a word that has been used to describe the continual diffusion of energy, which has nothing to do the concept of matter because matter can NEVER be created or **destroyed**...I think. (FYI, matter CAN be **destroyed**.) Or in simpler terms, entropy is the reason why your pheromonic perfume is able attract a sexual mate, regardless of their desirability.

Scientists have noticed that energetic quanta, which are NOT considered to be matter, are ALWAYS moving to regions of the universe that contain less energetic quanta. (All of which makes absolutely no sense UNLESS certain branches of the universe contain a substantial amount of positive or negative energy.) For example, the energetic quanta in a boiling pot of water will dissipate over time if the pot is removed from the heating source...unless the room is just as hot as the pot. Granted, scientists have postulated that heat-energy is ONLY stored about electron velocity within heated substances, but I don't believe such nonsense. All of which, is the reason why I postulated the existence of thermal energetic quanta, which are negative degradations of electrons. (Hopefully, it is only a matter of time before the scientific community embraces this logic and finds experimental data to support my postulate because it makes my unified theory so much more beautiful and logical.) And when this postulate is finally common knowledge within the scientific community, people will begin to wonder something even more profound: Is entropy ABSOULTELY dependent on the charge disparity within a universal branch? For example, is negative energy MORE entropic in super concentrated negative environments? Or, is negative energy LESS entropic around positive masses? Or, is space and time simply an abstract distance related entropic event that is dependent on our celestial micro-environmental battery system? Any who, hopefully these questions will enlighten you to the fact that there is an entropic spectrum. (Surprise, surprise!) Wait, I guess it is only surprising if you haven't read any of my other books, which is probably the case since I haven't sold any books. But, let's imagine that you've been paying attention and you're not surprised by my infatuation with a universe that exists as a collection of **spectrums**.

The rate of entropy in our branch of the universe, as it relates to non-matter energetic quanta, seems to be dependent on the negative density disparity. (Wishful thought: For those of you who are reading this for the second or third time, here is a caveat: MATTER distorts negative energy diffusion through the Admiral Ackbar and other methods.) For example, the **rate** of entropy is much greater within the heart of a campfire, as compared to the edges of the campfire, which is toasting your marshmallows, because of the density of negative energy within the heart of the campfire is much greater. Therefore, entropy is NOT a one size-fits-all sort of theory. Hopefully that brief overview of entropy will help you in later sections, if it hasn't dissipated from your mind. But for now, let's wonder onto a gravitational recap.

Previously, scientists thought that gravity was a mysterious force, which is exactly the **same** in EVERY part of the universe. All of which, is the reason I decided to postulate a mechanistic basis for gravity that depends on the positivity of an atom, charge disparity within a branch of the universe, the container-ish nature of the atomic orbitals, electromagnetic alignments, and the ability of disrupted atomic orbitals to shift atoms towards the negative gravitational energetic quanta to reinstate a symmetric atomic orbital arrangement. (Spoiler Alert: Gravity gets even more complicated and interesting in my future books.) In any event, the short-n-skinny is that entropy and gravity are the same thing: Diffusing negative energy.

Having all this in mind, it is easy to ponder the following: Is part of the Moon's gravity towards Earth the result of negative sunlight quanta reflecting off the Moon's surface and heading towards Earth? If so, then is this a revolutionary basis to quanta gravity? (Wait, maybe that extrapolative postulate was a bit of a jump for you'all. Let me slow down and explain that with a lot of complex words and then with some pornographic words.)

If certain atomic orbitals prefer to contain certain types of negative energy and certain types of atomic orbitals reflect/refract specific types of negative energy, then the association of atomic orbitals, i.e. quanta gravity, is NOT as clear as we'd like it to be. For example, let's say the atomic orbitals of atom #1 refracts negative blue photons into the atomic orbitals of atom #2, which degrades negative blue photons to produce a negative thermal energetic quanta, e.g. TEQ stick figure. Then, the negative TEQ stick figure diffuses back into the atomic orbital of atom #1, which is represented by the dotted arrow.

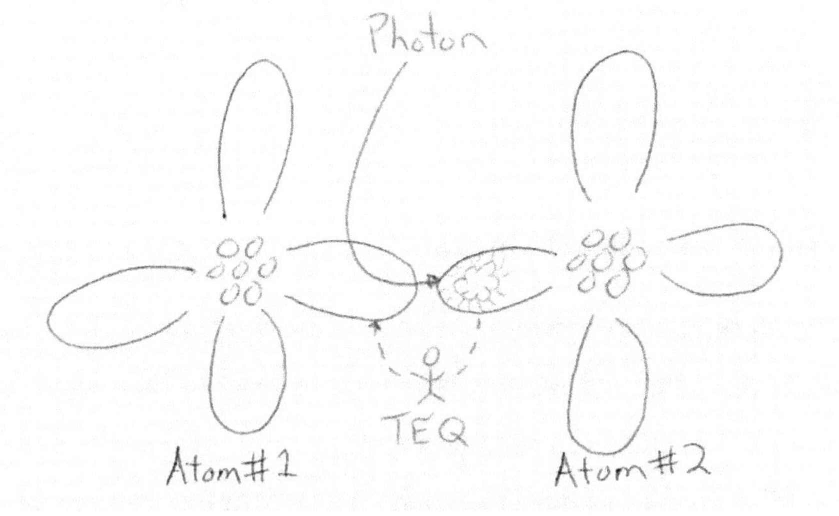


Figure 1: Gravitational Reflection/Refraction

As a result of atom #1 containing an atomic orbital that is expanding towards the TEQ stick figure that is being released by atom #2, atom #1 will move closer to atom #2 to maintain a symmetric atomic orbital arrangement, which is depicted by the arrow under atom #1 in Figure 2.

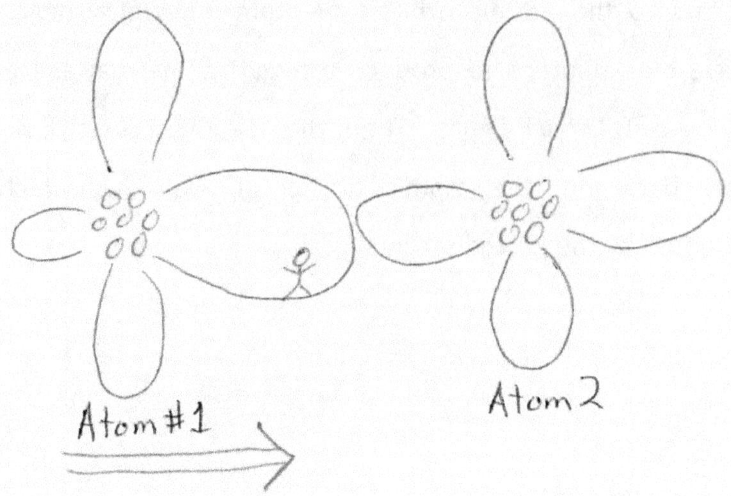


Figure 2: Atomic movement towards source of negative energy.

Therefore, quanta gravity is ALL about the ***direction*** of diffusing negative energy and HOW adjacent atoms react to that negative energy. Hopefully you see the simplicity of this example and the MASSIVE complexity it bestows upon the THEORY of quanta gravity as it pertains to a **spectrum** of negative quanta, elements, quanta degradation facilitated by atomic orbitals, and atomic orbital quanta reflection/refraction. Not to mention the massive complexity that it bestows upon the theory of macro-quanta gravity and celestial gravity. All of which, brings us back to the point at hand: The Moon is slowly moving towards Earth! (Ahhhhh! Help!?) Wait, if we colonize the Moon and cover it with solar panels, then the Moon will reflect less light from the Sun towards the Earth, which will decrease the gravity between these two objects and inhibit the impending collision. (Granted, the gravitational attraction between the Earth and the Moon is NOT solely based upon the Sunlight reflecting from the Moon towards the Earth, but that is beside the point. The point is that the negative thermal energetic quanta that are causing Global Warming will STOP the Moon from crashing into the Earth...I think?) In any event, science may be able to help society in the future. Who would have ever thought?

Instead of imagining that atomic nuclei are collections of super accelerated energy that are collected into protons and neutrons, which are exchanging positrons, let's imagine atomic nuclei as beating hearts. Also, if we completely disregard the color of blood based upon oxygenation, then we can imagine that the atomic nuclei hearts are red and the electrons coursing through the red atomic hearts are blue, which is easier to imagine since it mimics our circulatory system.

Figure 3: Atomic Hearts

As a result of the blue electrons squishing and squirting in, out, and about the red atomic nuclei heart, the red atomic nuclei heart doesn't ALWAYS have the blue negative life blood that it craves. Therefore, blue negative energetic quanta, like light or plasma, are always TRYING to squish past the BLUE electrons towards the positive RED atomic nuclei heart. Fortunately or unfortunately, depending on your relative emotional connection with the red atomic nuclei hearts and/or the blue negative plasma, the blue negative plasma squishes towards the red positive atomic nuclei heart, but the blue negative electrons squirt the blue negative plasma away from the red positive atomic nuclei heart.

Figure 4: Squishing and a Squirting

So even though you are probably uncomfortable with all this pornographic blue negative plasma that exists about atoms, it is imperative to the atomic battery system, from which we are forged.

Another way to understand the above mentioned squishing, squirting, and Quanta Dynamics, is to imagine that atoms are living and breathing entities. As atoms lose/gain electrons and/or as electrons degrade, atoms beat and breathe their electrons differently based upon the different negative energetic quanta surrounding the atomic nuclei heart and the atomic nuclei heart's need for negative energetic quanta to sate the positive charges within the atomic nuclei heart. (If you have trouble imagining WHY electrons slowly degrade and lose negative energy, just remember atomic nuclei are always beating the poor electrons to death to protect the positive energy within the atomic nuclei because negative energy is so pervasive in our negative branch of the universe.) Quite simply, it is a cold harsh universe out there and if elements didn't associate with as many different elements as possible, even though they may have different rhythms of squishing and squirting, then the universe wouldn't have collections of matter sticking together. Without external negative energy for atoms to SHARE, atoms would shrink, degrade, and die, which would completely negate the whole battery system existence that we are currently enjoying.

In conclusion, I'm afraid of the Moon crashing into the Earth, there is an entropic spectrum, and that quanta-gravity is a lot like celestial-gravity, except for all the different factors and stuff like that. And finally, I'm afraid we exist in a magnificently complex squishing and squirting orgasmic galactic battery system, which makes me horny enough to squirt out some DNA.

# Chapter 3

I know what most of you are thinking: A futuristic tractor beam seems highly unlikely, given my gravity postulates. Actually, a tractor beam would be really difficult to invent if someone invented a negative force field, which would reflect/refract the negative energetic quanta of the 'tractor beam'. In any event, now seems like a perfect time to ponder all that futuristic science fiction stuff. Not because it actually fits into the flow of this book, but because the previous chapter was way too technical to be interesting...I think.

Since a portion of quanta gravity is the result of the atomic orbital modification and the subsequent movement of the atom via the movement of its atomic orbitals to restore some level of atomic orbital symmetry, the idea of a tractor beam seems absolutely stupid. Granted, OLD tractor beam postulates imagined that gravity was a "pulling" force, but that is completely illogical in a quanta world. Therefore, let's break free of the old verbal-shackles and state the following: Gravitational energy makes atoms ***PUSH*** toward the source of negative gravitational energy. Unfortunately, there is one major problem with the concept of a tractor beam and the conceptualization of negative gravitational energy: The energy within the tractor beam will degrade into thermal energetic quanta, which will cause the object to MELT long before the object in the tractor beam begins to **PUSH** itself in the direction of the tractor beam. Not only will the gravitational energy degrade into thermal energetic quanta, the degraded gravitational energy will also cause the reflection/refraction of energy in the 'tractor beam', which will decrease the directional specificity and intensity of the tractor beam energy. And finally, the negativity of the gravitational energy will modify the positivity of the object in the tractor beam, which will change the rate that the object PUSHES towards the gravitational energy.

If one were able to invent a tractor beam, what would be some beneficial characteristics to a tractor beam? First and foremost, one could have hillbilly tractor-pulls in space. Next, a tractor beam should adjust the environment around the object such that NO other negative energy is innervating the desired object, which might be acting against the ability of the tractor beam to cause the object to PUSH in the direction of the tractor beam. Then you could make a ton of money selling illegal moonshine to hillbillies in space as they hoop-n-holler for their favorite tractor-beam. Next, since there are multiple spectrums of energy, atomic orbitals, and degradation products, tractor beam energy must be specifically designed per the object to maximize the specific elements PUSH-ability. In addition to all that, the variable negative energy of the tractor beam must re-orientate the atomic orbitals in the desired object while maintaining a **flush** of negative energy around the desired object such that the overall positivity of the object IN the tractor beam is NOT altered. And finally, the gravitational energy cannot modulate the entropy of the degrading gravitational energy, which could result in the object melting within the tractor beam. (So much for being less technical...sorry.) In short, the composition of the tractor-beam energetic quanta must be modified such that the positivity of the object and the refractivity of the object is accounted for in order to create the movement of the object as a result of the objects adjusted atomic orbitals. THEORETICALLY, this is exactly what the Sun is doing to the Earth...and the other planets. Wait, I just had a brain storm...hopefully it is less technical.

As a result of atomic orbital saturation from incoming negative gravitational energy, which could decrease the PUSHING of atoms towards the negative gravitational energy, celestial gravity only works on ROTATING objects. (Wow, I think I just blew my own mind.) Granted, the rate of the gravitational-rotation crux is absolutely dependent on the type of negative energy in the gravitational energy, the type of matter in the object, the intensity of the negative gravitational energy, and the presence and composition of the area around the object, but that brain-storm is totally off the hizzy! I mean, that is totally whack man!

If a celestial object is NOT rotating, then the atomic orbitals facing the negative gravitational energy will become overwhelmed with the negative degradations of the gravitational energy, which will refract incoming gravitational energy. As a result of refracting incoming gravitational energy, less negative energy is present to cause the atomic orbitals to push TOWARDS the negative gravitational energy. All of which means, rotation of the object MAXIMIZES the **diffusion** of negative gravitational energy degradations of from the object's atomic orbitals and MAXIMIZES the force of the atomic orbitals **pushing** towards the negative gravitational energy.

Or in simpler terms, just imagine that the Earth and the Sun are two boxers. Next, imagine that the Sun continually punches the Earth with negative photons. And finally, imagine the Earth takes to photons, degrades them into negative thermal energetic quanta, and begins punching back at the Sun with negative thermal energetic quanta. As soon as the Earth begins to punch back with negative thermal energetic quanta, the Earth begins to spin and the Sun begins to punch the other side of the Earth, which doesn't have enough negative thermal energetic quanta to punch back.

In conclusion, pondering the concept of a tractor beam has totally opened up another factor to celestial gravity, which kind of fits into the flow of this book. Or in more complex terms, I postulate that gravity is based upon the ability of atoms to PUSH towards negative gravitational energy in this branch of the universe. Also, I postulated that rotating celestial bodies maximize the diffusion of degraded negative gravitational energy such that an atom's atomic orbitals do not become oversaturated with negative gravitational energy and stop PUSHING towards the object irradiating the negative gravitational energy. All of which, maximizes the gravitational force created by the atoms pushing towards the source of the negative gravitational energy. Granted, there are tons of factors to this undulating postulate, but I need to maintain a constant theoretical rotation otherwise you'll stop reading because your brain will become over-saturated and begin to reflect/refract all my ideas...I think.

# Chapter 4

Of course I have no clue what I'm talking about...wait. I didn't mean to write that. What I meant to write is: The underlying logic of 'gravity' causing objects to PUSH towards negative gravitational energy is based upon several postulates. So for your pleasure, let's review some of those postulates. (Actually, there is no logical reason why I should be reviewing things at this point in the book. Apparently, I'm trying MINIMIZE any interest you might have in learning this complex shit...I think.)

First and foremost, I postulated that negative electrons are recharged by traversing the positive atomic nucleus. (Imagine a red atomic heart pounding out blue electrons.)

Next, I postulated that positive charge dispersion caused by the exchange of positrons within Neuproz groupings causes a unique and delocalized positive charge surge about atomic nucleus, which is dependent on the arrangement of ALL the Neuproz groupings within the atomic nucleus. (Imagine that Neuproz groupings are just multiple Neuproz pairs, which are playing hot-potato with all the positrons.)

Then, I postulated that atomic orbitals like to maintain their symmetry about the atomic nucleus because this allows for less electrostatic repulsion between the negative electrons. (Just imagine that electrons do a better job at keeping crazy negative plasma away from the positrons when electrons are arranged symmetrically.)

Next, I postulated that positrons are moving much faster than electrons and will degrade when combined with too much negative plasma and/or an electron. (Just imagine an atomic explosion.)

Then I postulated that electrons create Energy Containment Regions, e.g. atomic oribtals. (Just imagine that atomic orbitals are like little bitty coffee cups that contain lots of hot negative stuff.)

Next, I postulated that electrons do NOT degrade perfect packets of quanta energy unless they are stimulated to **degrade** by running through a dense cloud of negative plasma. (Just imagine a negative duck crashing into an electron-airplane-engine, which causes the electron-airplane-engine to explode and release a bit of energy in the form of light or magnetic energetic quanta.)

And of course, I postulated that the extreme presence of negative energetic quanta about an atom can cause the distortion and/or the destruction of atomic orbitals and the ionization of the atom, which is the basis as to why Earth's core is releasing the electrons that are causing lightning. (Just imagine that I didn't just postulate that.)

And finally, I postulated that the number and arrangement of Neuproz groupings in the atomic nucleus and subsequently, the movement of the all the positrons dictates the movement of the electrons, which determines the characteristics of the atomic orbitals that are innervating the external environment by destroying, containing, or deflecting external negative energy that is unique to our branch of the universal energy tree. (Just imagine that English is my second language and I don't know anything about run-on sentences.)

Unfortunately, that was a very limited overview designed to align you with my next postulate: Your mama's fat and she dresses you funny. (Gravitational SLAM…when she falls!) I've been waiting my whole life to use that insult…I think. In any event, after that melodramatic unimportant gravitational insult that was refracted around your mama's fat gravitational capacity and directed at you, one question still remains: With all those layers upon layers of complexity engulfed by a spectrum of Entropic rates and factors, is it possible to create a tractor-beam? The short answer is: Sure. The long answer is: Gravity. And another answer is: This should have been in the previous chapter.

You might not agree that the long answer is longer, but it does have more letters and it does contain a massively complex postulative-lineage. In any event, gravity is a tractor-beam that works quite well on a celestial time-scale, but not so much on a melodramatic-science-fiction time-scale. I mean, most people would probably change the channel if it took three weeks for a Starfleet tractor-beam to pull in a rebel space-craft.

In conclusion, it might be possible for someone to discover a mixture of energetic quanta that would act as a tractor-beam and cause an object to PUSH itself towards the negative energetic quanta without melting the object, but I have a feeling that Anti-gravity devices will be easier to make and thus, will be more sophisticated by that theoretical point in human history when tractor-beams might be invented. All of which, depends on the course of human history, which seems to be stuck in an Intellectual Bermuda Triangle worm-hole that is connected to the dark ages. (BTW, I only believe in mental-wormholes.)

# Chapter 5

Is that George Clooney over there?!!!

Ok, now that it is just us dudes, it is time to tackle this grossly inaccurate gravity kerfuffle, which has nothing to do with that movie George Clooney was in because gravity exists as a spectrum. Quite simply, even though almost every element can exist in a POSITIVE oxidation state, aside from those asshole noble gasses that are too proud to share their electrons, nobody has taken the time to think about planetary oxidation states and how they correlate to celestial gravity. Therefore, I'm going to talk about celestial and quanta gravity, which are at the opposite ends of the gravitational spectrum…I think.

A long time ago in a galaxy called the Milky Way, I made the astute observation that the gravity measured on each planet in our solar system did NOT correlate to the mass of each planet. Granted, I was just another asshole making poignant observations as it correlates to the insides of my brain, but now I feel a tad-bit more comfortable in stating the following: There is something fishy about this gravity-hoopla and its over-reaching 'universal' constant. All of which, I will begin to discuss, albeit in the next paragraph.

Maybe it is a consequence that we reside on a branch of the universe in which the fundamental efficiency of the matter battery systems, for which we are made of, are forged from charge separation. Or maybe, our branch of God is held together by positive atomic nuclei that are continually 'swinging' with lots of Devil electrons. Or maybe, the positive charge movement in atomic nuclei dictates the arrangement of the atomic orbitals, which dictate elements' ability to have positive oxidation states. In any event, an even odder thought should come to mind: Even though a large portion of positively charged atoms can exist without a few negative electrons, nobody has extrapolated this charge kerfuffle to planets and/or celestial gravity. And on top of all that hoopla, my postulate that thermal, magnetic, and electromagnetic energetic quanta are NEGATIVE degradations of electrons gives a mildly amusing mechanistic description of 'gravity' as it pertains to positivity in our solar system. Of course, I need to take a moment and smooth out the edges of this postulate with a few more postulates, but I think you'll find one of my postulates amusing...at some point.

Everybody knows that the composition of Earth is different than the other planets in our solar system. (By everybody, I mean most scientists.) In addition to that, everybody knows that asteroids contain elements that are NOT common on Earth. All of which, leads me to several postulates:

1. As a result of Earth's unique negative magnetic energetic quanta degrading in the atomic orbitals of the elements on Earth, Earth modulates the half-lives of the elements on Earth. Or in simpler terms, Earth is giving out tax-breaks to the worthy elements to make sure they last longer.
2. The specific and continual selection for elements on a planet can select for a uniquely positive planet.
3. The unique positivity of a planet not only modulates a planet's elemental concentration via the unique magnetosphere, but the planet's positivity will also determine how a planet reacts to a star's negative energy, e.g. celestial gravity.

So what is the take home message? Little Joey is failing science because he's more interested in girls? On second thought, the take home message should be: Things are complicated. For example, let's say that Earth's 'supposed' iron core is made up of iron with a +2 charge, but in a couple hundred million years the continual degradation and diffusion of negative energy will result in Earth's core evolving into iron with a +3 charge. Granted, the celestial gravity will be different because the Earth will be more attracted to the Sun's negative energy, but you also have to factor in the electromagnetic component of celestial gravity. In simpler terms, a more positive Earth core will mean there is less negative energy to cause the stimulated degradation of electrons to produce negative magnetic energetic quanta, which are aligning Earth to the Sun like an iron filing around a magnet. All of which means, the evolutionary degradation of a planet's matter will modulate the planets oxidation state, which will modify how the planet is attracted to a nearby star, but the evolution of the planet's matter and oxidation state will also modify the release of magnetic energetic quanta by the planet. Unfortunately, all of this is complicated by the Sun.

I don't know if you recall this, but I postulated that atomic orbitals can wobble and have a tendency to be influenced by the external environment. Therefore, a young planet, and subsequently the elements therein, can be influenced by the Sun. For example, the placement of a young planet around a star will result in the selection of certain elements, which will eventually create a magnetic environment around the planet that selects for the position around the star. All of which means, the evolution of a planetary oxidation state is correlated to the Sun's effect on any elemental mass that gets too close to the Sun.

In preemptive conclusion, it seems to be apparent that the Sun is influencing HOW the Earth's quanta gravity evolves and results in celestial gravity. Now all I have to do is stretch this book out a little further. So, here is some more stuff on oxidation.

Although it might be perceived as a conceptual energetic flaw for elements to be stable in different oxidation states, there are two distinct benefits to the oxidation of battery systems.

1. Atomic orbital wobble results in stimulated electron decay to produce magnetic energetic quanta, which facilitates the cohesion of stable matter batteries via quanta gravity.
2. Positive elements are attracted to the negative energy of stars, which facilitates the cohesion of celestial batteries via celestial gravity.

All of which, is called GRAVITY and allows for the continual association of matter into battery systems. Or in more complex terms, the positive oxidation states of elements in the periodic table allow for the formation of efficient solar battery systems, which are components of the galactic battery system, which are components in the universal-branch battery system, which is a component in the universal battery system. Therefore, the life of a galaxy is intricately based upon elemental oxidation states. Once enough elements have decayed into neutral species, the galaxy will begin to expand exponentially.

With all that in mind, why are the celestial and galactic battery systems moving spirally? Well, positrons are moving within atomic nuclei to avert electrons from finding the positive charge; Electrons are moving to protect the atomic nuclei from negative plasma. Therefore, celestial and galactic battery systems are moving to protect stars from asteroids and/or external energy, which will decrease the battery capacity of the solar system and/or galaxy. So what is the purpose of planets? Well, unfortunately for Earth and all the species there in, planets are a secondary defense against asteroids. Or in simpler terms, planets are really good at defecting or absorbing asteroids, which is great for the life of a star, but NOT-so-great for the life on the planets. So even though a star protects itself for asteroids by constantly swirling and releasing large amounts of negative energy, spinning planets can also deflect asteroids and provide ANOTHER layer of protection for stars.

So technically, we exist on a piece of a massive battery system in which Earth's only purpose is to be a secondary star defense. As for the reason why our planet is made of mostly oxidized metals, well there are three intertwined reasons:

1. Oxidized positive elements have the ability to continually adjust their atomic orbitals and push towards the negative energy that is diffusing from stars.
2. Oxidized positive elements have the ability to adjust their atomic orbitals, e.g. atomic orbital wobble, to facilitate the stimulated decay of electrons, which yields magnetic energetic quanta. Magnetic energetic quanta is a component in celestial gravity and also does a great job at creating Earth's wonderfully protective magnetosphere.
3. Oxidized positive elements decrease the rate of entropy of negative energetic quanta, which increases the longevity of matter batteries. Also, the exchange of negative energetic quanta between oxidized positive elements causes the associate of oxidized positive elements, which form planets and protect solar battery systems.

In conclusion, our galaxy is an efficient battery system, which has many intercalated layers. First, the Sun facilitates the specific, but slow, decay of matter on the planets in the solar system. The specific and slow decay of a planet’s unique concentration of matter determines the planet’s magnetosphere and positivity, which dictates the planet's orbit. And finally, planets' orbits are designed to protect the Star from being continually sand-blasted by all the matter in our branch of the universe. All of which is great for our Sun, but not so great for the things living on any given planet.

# Chapter 6

The only question that remains, is nothing but a trivial quandary: Do oxidized elements have longer or shorter half-lives? On one hand, oxidized elements tend to associate with the negative stars. On the other hand, neural elements don't have to deal with as much CRAZY negative plasma, which leads to less electron degradation. Unfortunately, as with many things in science, this quandary can NOT be answered by obtaining data from ONE planet around ONE star in ONE galaxy in ONE branch of the universe. In any event, I think we should let Mr. Logic say something first before I answer the preceding question with a 'maybe'.

"Let me see." Mr. Logic begins with a British accent. "Elemental oxidation results in stronger columbic repulsion within the atomic nucleus, but the basis of Battery Unified Theory Theory (BUTT) dictates that elements have longer half-lives when contained within larger battery systems. And for the record, quirks don't twerk, but they can be twerps."

In light of Mr. Logic's stimulated emission about BUTT and half-lives, it is important to note that life has not changed. People still do not believe in Mr. Logic or his BUTT. People still don't believe in the existence of NEGATIVE magnetic, electromagnetic, or thermal energetic quanta. People don't believe in evolution. And finally, people don't believe in energy efficiency. Actually, if energetic quanta were people, they would ONLY believe in energy efficiency because of its life-sustaining ability, but would still argue against an economy that is based upon energy efficiency. (Apparently, most energetic quanta lack the vision of a cohesive atomic nuclei and think most electrons would be lazy and hedonistic if they were incorporated into an energy efficient economy, but that is beside the point.)

Having said all that, I need to make an impassioned plea to American society: For the love of God, be a little more mindful of your place in the world. I know that Americans can ONLY serve their country by joining the military, but non-intellectualism is the BIGGEST terrorist with respect to **American Ideals**. The second biggest terrorist, with respect to American Ideals, is the evening news. And the third biggest terrorist, with respect to American Ideals, is wasteful consumerism. If you think that saving a buck by NOT buying energy efficient appliances, cars, or homes is justified by the sub-standard food you will consume with those savings, then maybe it is time America had an intervention.

"America, please come in and sit down." Counselor

"What the hell is going on here?" America

"Don't be angry. We are ALL your friends here." Counselor

"Bullshit, I hate fucking Iran!" America

"Granted, Iran has its own problems, but this is about you America." Counselor

"About me? Bullshit!" America says as it begins to scratch its oil drums. "Iran wants to build a bomb and destroy Israel."

"Please do not deflect your own problems by getting angry at Iran. They are working on their problems, but this is about you America." Counselor

"What about Benghazi, Syria, and gay rights in Russia? That shit is fucked up!" America yells as it scratches a hole in one of its oil drums, which begins to ooze.

"Look at yourself America. You haven't slept in years, you think everyone is trying to kill you, and you're always throwing your money at other countries' oil reserves. Maybe it is time you decided to turn on the light and face yourself in the mirror." Counselor

"Fuck you! I don't have a problem. They have a problem and I'm the only sane one in this room." America

"Come on America, you know that isn't true. What about Great Britain?" Counselor

"Fine, Great Britain isn't that crazy. But the rest of you are fucking bonkers." America

"That is a good start America." Counselor

"Listen man, all a need is one little pipeline from Canada and I'll be fine." America

"No America. Canada is NOT America and Canada deserves to use their own resources for the development of their country." Counselor

"Fucking Canada doesn't even like oil, right Canada? They are just dying to give that shit away for a shot at the American Dream...Right Canada?" America

"Not everybody wants to be America." Counselor

"Why not?! It is fucking great to be America!" America

“What have you accomplished lately?” Counselor

“Well there was that works project I did.” America

“You never followed through with it.” Counselor

“What about that healthcare reform.” America

“You’re still the ONLY industrialized nation that does NOT provide healthcare for ALL of its citizens.” Counselor

“I went to the fucking Moon, you remember that! That is a fucking accomplishment.” America

“That was a long time ago. Now you’re so addicted to foreign oil that Russia has to send your astronauts to the international space station.” Counselor

“Can I say something?” France says with an accent.

“Fuck you France!” America

“No this is good. What do you have to say France?” Counselor

“We helped you with the American Revolution and how do you repay us? Sub-prime mortgage economy collapse!” France

“Yeah!” Several Countries

"Fine, I'll put up some windmills like Belgium." America says with a whimper.

"You need to cut your oil addition cold turkey!" Iran

"Fuck you and your massive oil reserve. I will come over there and fuck you up! You hear me?!" America

"Calm down America. Nobody has the energy to waste over petty squabbles." Counselor

"I'm calm...it is just that..." America

"It is just what? You can say it here, this is a safe place." Counselor

"Well, if I go oil-cold-turkey, nobody will admire me for my stupid fuel efficient cars." America

"Do muscle cars define you?" Counselor

"Yes! I love the way the car rumbles underneath me and every country turns to look at me when I drive by." America

"No, muscle cars do not define you America. What is in your heart defines you. Muscle cars are just something you use as a distraction because you are extremely insecure." Counselor

"My Florida hasn't been working lately." America says with a mumble.

"And why is that?" Counselor

"I don't know, voter fraud?" America said as a tear rolls down its cheeks. "Florida just sort of hangs there and reminds me about how great I use to be. I used to be strong and I didn't need anybody's help."

"You are still strong America, but your heart has been corrupted." Counselor

"If I just frack some shale, I'll be fine. I swear!" America

"Will fracking fix the problem America?" Counselor

"No." America says and begins to cry.

"What do you need to do?" Counselor

"I don't want to go GREEN, it sounds so gay! I want to drive a massive muscle car for the rest of my life until I crash or it explodes!" America says while weeping.

"You don't want a future?" Counselor

"No, I don't want to live anymore!" America

"Is that true?" Counselor

"No." America

“Do you want a future?” Counselor

“Yes.” America says and stops crying with a long sniff.

“If you want a future, then what do you have to do?” Counselor

“Go green.” America

“Do you mean it this time?” Counselor

“Yes.” America

“We’ve heard that before.” Iran

“I will fucking cut you to pieces with bombs!” America

“America?” Counselor

“Fine, I’m sorry.” America

“You know that if you don’t mean it this time, you will keep spiraling further and further down until there is no future for your children.” Counselor

“I know.” America

“This is your last chance. You need to dream of a new beginning that will provide a prosperous future for ALL your children.” Counselor

“I can.” America

“You can what?” Counselor

“I can change.” America

“Good.” Counselor

“Is it cheating if I drill off shore?” America

“You know the answer America.” Counselor

“I guess.” America

“If you drag your feet, you will only make things worse for everybody.” Counselor

“I know, but…” America

“But what?” Counselor

As you know, this intervention has been going on for a really-really long time and the arguments have been getting more and more convoluted. So where does this leave us?

In conclusion, although oxidized elements might seem to be less energy efficient, they are attracted to negative stars: By the emission of magnetic energetic quanta and by being positive. Therefore, oxidized elements have longer half-lives because they have a tendency to exist within negative celestial battery systems.

# Chapter 7

Even though the universe is energy and is designed around energy efficiency, it is a monumental battle to be energy efficient, educate people about energy efficiency, and inspire people to be energy efficient. Here are a few ways that America can improve its energy efficiency: Energy Reclamation; Laser spark plugs; Stronger magnets for electromagnetic induction. All of which, can be incorporated into automobiles. Therefore, let me be pedantic for a moment and apply most my theoretical postulates towards the automobile...Vroom-vroom honky-honk.

Negative plasma released by atomic orbital distortion creates a chemical chain reaction in your automobile engine. Therefore, when a spark plug initiates the oxidation of gasoline in your engine, it is the subsequent release of negative plasma by oxidation that creates the chain of reaction within your engine, which oxidizes the gasoline into smaller molecules that drives the pistons in your engine. Unfortunately, some of the smaller molecules produced by this oxidative reaction are not so good for humans and/or human existence. That is why scientists invented the catalytic converter, which uses heavy metal catalysts and heat to ensure that most of the molecule leaving your car's engine is oxidized to carbon dioxide.

With all that in mind, here is how I would use my postulates to improve the efficiency of automobiles:

1. The use of laser spark plugs will be an exponential improvement to the RATE of the oxidative chain reaction because light moves FASTER than the diffusion of thermal energetic quanta.
2. An increase in the RATE of the oxidation chain reaction will increase the amount of oxidized gasoline, which will result in greater POWER via the production of more $CO_2$. Or in simpler terms, more $CO_2$ to move the pistons in your automobile engine.
3. With more POWER per unit of gasoline, less gasoline will be needed to provide the same amount of POWER.
4. With less gasoline required for the same amount of power, less heat (Thermal Energetic Quanta) will be released during combustion.
5. Since current engines are designed to run at an optimal temperature, which is a temperature that facilitates the

oxidation chain reaction, the excess thermal energy can be reclaimed.

6. Reclaimed thermal energy can convert water into steam, which can push a turbine to create energy via electromagnetic induction.
7. Formation of stronger magnets can improve the efficiency of electromagnetic induction, which will allow more energy to be reclaimed from the thermal energy released by the oxidative chain reaction.

As you can see, each layer of the combustion engine has been designed to function in the presence of every other engine component. For example, HEAT (Thermal Energetic Quanta) facilitates the oxidative chain reaction in your engine. Therefore, engines are designed to maintain heat between the required catalytic amount of heat and the amount heat that will fracture your engine block. But for the sake of pedantic, let's imagine a perfect engine.

Let's imagine a silver-iron engine block that is cold to the touch because all the heat is trapped in the combustion chamber. Next, let's imagine that a valve system has been invented that allows for laser spark-plugs. And finally, let's imagine the heat, which is NORMALLY used by the catalytic converter, converts water to steam, which will turn a tiny turbine and produce electricity. With all this in your imagination, it should be simple to imagine that this engine would be more energy efficient because:

1. Excess thermal energetic quanta (heat) are NOT thrown away via the traditional radiator system.
2. The RATE of the oxidative chain reaction will be enhanced via the movement of light instead of the diffusion of heat.
3. With an enhanced oxidation rate, less gasoline will be required to produce the same amount of force on the piston.
4. Excess heat can reclaimed by converting water into steam via pushing a tiny turbine, which creates electricity via electromagnetic induction.
5. Electromagnetic induction is used to slow the car down while producing energy.

Unfortunately OR fortunately, Honda and other car companies have already invented the ability of cars to brake while recharging their batteries. But, in my imagination, the efficiency of this conversion is more efficient because someone has invented stronger magnets. Therefore, let's take a moment and investigate my imagination...Oh the HORROR! Wait, wrong imaginary sequence.

In my imagination, the energy efficiency of electromagnetic induction is really lousy because it is ALL based upon Quantum Mechanics instead of Quanta Dynamics, which gives a method to the stimulated degradation of magnetic energetic quanta and subsequently the basis of electromagnetic induction. But that is all beside the point. The point is: Stronger 'cheaper' magnets will make electromagnetic induction more efficient. Therefore, here are a couple ideas about increasing the strength of magnets:

1. Increasing the density of the magnets will decrease the diffusion rate of thermal energetic quanta, which are the energetic quanta that are stimulating the electrons in iron to decay and release magnetic energetic quanta.
2. Increasing the density of the magnets will increase the atomic orbital wobble, which causes electrons to run into thermal energetic quanta.
3. Increasing the density of the magnets will result in greater force applied between iron atoms, which result in a denser electromagnetic field. (FYI, 'force' is being used to describe the transmittance of magnetic energetic quanta between iron atoms.)

Hopefully, somebody is confused as to how atom density is correlated to electromagnetic field density. If not, I'm going to explain it anyways.

Everybody knows that when you sprinkle iron filings around a magnet, the iron filing align to form these 'flux' lines. Now what those 'flux' lines represent are the linear superposition of ALL the magnetic energetic quanta being released by the magnet. Or in simpler terms, not ALL magnetic energetic quanta are being released from the absolute 'north' pole of a magnet. Therefore, the MORE iron atoms are aligned within the magnet, the MORE magnetic energetic quanta will be focused to be released from the absolute 'north' pole of a magnet. The denser the magnetic energetic quanta are at the absolute 'north' pole of a magnet, the more these magnetic energetic quanta will be able to distort the electromagnetic micro-environment within the wire around the magnet, which causes electrons to move. The more electrons move per the movement of a magnet, the more efficient electromagnetic induction becomes.

In conclusion, a lot can be done to improve the efficiency of internal combustion engines and all the factors therein, but it all starts with understanding Quanta Dynamics. Also, it might be helpful to understand Organic Chemistry, but I think that is probably asking too much.

# Chapter 8

In as much as I'd love to wrap this book up with some poignant remarks and move onto my next book attempt, I feel obliged to share something I've been thinking about: Interstellar Travel.

Currently, it is impossible for humanity to make stable matter undergo fission. Maybe we haven't created strong enough lasers or found the right combination of energy that will facilitate normal matter to degrade by fission. But, there will be a day when scientist, working outside the Earth's atmosphere, will discover a way to cause normal matter to degrade by fission and this will facilitate interstellar travel.

For those of you who are unfamiliar with propulsion, you need energized matter to shoot out the back of your spacecraft to make it go forward. Therefore, interstellar propulsion becomes a problem of **preparing** and *storing* the energized <u>matter</u> that "squirts" out the back of your spacecraft. (Preferably, matter should CONSISTANTLY squirts out your spacecraft's butt so that the interstellar travel isn't lurchy.) Butt, what if you could pick up any old ROCK and make it into a fission ROCKet? Well, astronauts won't have to worry so much about the purification and storage of propulsion material, which will leave a lot more room for food and oxygen on the spacecraft.

Granted, initial prototypes of this ROCKet system will be lurchy and possibly explosive, but I have no doubt that scientists will be able to make educated guess about the composition of rocks as well as the energy needed to turn unique rocks into ROCKet propulsions systems.

In conclusion, although repeatability doesn't really facilitate a logical theory as it pertain to the beginning of the universe, repeatability will allow scientists to create a ROCKet system that will enable humanity to explore the universe...hopefully.

# Cliff Notes

1. We are human and we can't see everything. Therefore, we cannot see the whole universe.
2. The universe is made of energy. Therefore, the conservation of energy dictates that the universe consists of intercalated battery systems.
3. Energy degrades and ALL energy fragments are NOT the same. Therefore, the universe is a collection of different energy fragments or branches.
4. Since each universal branch is different, each universal branch contains a unique amount of positivity or negativity.
5. Protons are really large in this branch of the universe. Therefore, our branch of the universe is overwhelmingly negative.
6. Gravity and Entropy are the same thing in as much as Gravity exists as a **result of** negative energy diffusion and Entropy exists **because of** negative energy diffusion in a negative branch of the universe.
7. Science is the study or reproducible events. Therefore, it is only logical that scientists postulated that the Big Bang will repeat. Unfortunately, this is wrong since Gravity and Entropy are both based upon negative energy diffusion in this branch of the universe.
8. Negative electrons protect positive atomic nuclei by corralling the negative plasma into atomic orbitals or by squirting the negative plasma away from the atomic nuclei.
9. Electrons have more negative components than positive components.
10. When a positive component of an electron is destroyed by the collision with an external negative quanta of energy, the electron

compensates for this loss by emitting a negative photon or magnetic energetic quanta.

11. Quanta gravity is directionally specific movement of negative energy from one atom towards another atom via diffusion, reflection, or refraction.
12. Celestial gravity is a collection of magnetism, planetary oxidation states, and the diffusion of negative energy that is maximized when an object is rotating to avoid surface saturation with negative energy and subsequently the reflection/refraction of the negative energy of Celestial gravity.
13. As a result of the Battery Unified Theory Theory (BUTT), it has come to my attention that Gravity, positive density, and the proximity to OTHER branches of the universe are the **main** factors in the RATE of entropy. Therefore, there is a spectrum of entropy based upon one's location within a given branch of the universe.

# II. The Conformational Flow

While at college, I went on a white water rafting trip, which was sponsored by the recreation center. And much to my dismay, I did not find a spunky outdoorsy mate that was obsessed, in a good way, with reclusive want-a-be intellectuals. What I did discover was that rivers are not straight. That is not to say I didn't already know that, but idea was jarred into my head by the white water rapids. All of which, has nothing to do with this book...I think. I mean, we are made of about 70% water and our bodies are based upon multiple water-flowing systems, like the cardiovascular and lymphoid systems, but that is beside the point. The point is, the human nervous system runs on ELECTRICITY because Frankenstein was reanimated with electricity. Therefore, this book is ONLY about my white water rafting trip and NOT about my inability to change the world through procreation. On second thought, I think the nervous system would be more interesting.

In as much as it is easy to think of the lymphoid system as a set of canals, the cardiovascular system as a set of super highways, and the nervous system as a set of electrical lines, things are a little bit more complicated than those simple analogies. For starters, both the lymphoid and cardiovascular systems are designed to move FLUID about our multicellular body, but the nervous system is NOT designed to move any substance? Wait, the nervous system is a set of TUBES, just like the lymphoid and cardiovascular systems, but the nervous system runs on DUNKIN DONUTS? Wait, I think that was just an integrated advertisement and not a metaphor about the tubular nature of the nervous system. Here is what I meant to ask: Is the nervous system just a set of TUBES that are **NOT** designed to move any substance because the nervous system runs on electricity? Wait, does that make any sense?

What do the following phrases have in common: The spark of God; A mental lightning storm; Electrical impulse; A shocking thought? Well, they all reinforce the theoretical belief that your nervous system runs on electricity. Granted, there is plenty of 'supposed' empirical data to support the notion that your nerves operate by electricity, but there is one problem: The body's ULTIMATE function is to CONTROL unwanted **oxidation** and/or reduction reactions. Hopefully, I have sparked your interest...wait, damn it. I mean, I hope I've squished your interest. Aw, damn it. Never mind. Here are some more weird questions about the nervous system:

1. How does this random Reduction-Oxidation chemistry JUMP myelin sheath cells?
2. Why are nerve cells NOT designed like Mitochondria's electron transport train?
3. How does the ATP-energy get from the synaptic bulb to ALL the transmembrane ion channels along the neuron to repair this supposed 'electrical gradient'?
4. How does this RANDOM oxidation/reduction NOT denature/destroy the surrounding proteins?

Hopefully, you can see some MASSIVE problems in the OLD 'electrical' theory of the nervous system. Therefore, the ultimate question of this book is: Does the nervous system run on micro-fluid dynamics? Unfortunately, this question angers lots of grumpy old scientists and their entrenched scientific beliefs. Butt, I think my position, that the nervous system runs on micro-fluid dynamics, will be absolutely obvious to you by the end of this book...I think. If you're not completely satisfied, I will refund all the neurotransmitters that you've squirted out in anger, minus shipping and handling.*

*The author is not responsible for the force required to push synaptic vesicles towards any **angry** neurological synapses. All neurotransmitter refunds are based upon the author's basal line secretion of neurotransmitters, which has been modulated by continual ingestion of SSRI. Refunds are not available in Colorado, Washington, or Amsterdam.

# Chapter 2

I could have started this book with EVERY postulates of how the body works since the beginning of time, but that would seem counterproductive since I'm trying to postulate something new. Therefore, let me just say that the collective cognition of Homo sapiens as it pertains to **how** and why the body works the way the body works, has NOT always been precise or even accurate. As a species, we've had some pretty wild ideas as to HOW the body does what the body does. Hell, if we didn't have an evolution of thought every so often, we would still be having plagues.

To begin this new intellectual journey, let's imagine that the lymphoid system looks like a bunch canals. Actually, if you've ever studied anatomy-physiology, you'll know that the lymphoid system actually looks like a bunch of canals.

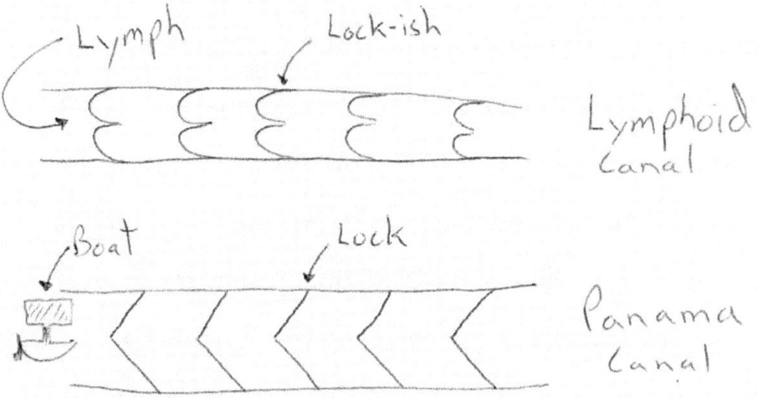


Figure 1: Lymphoid System vs. Panama Canal

As you can see in this figure, the Panama Canal and the lymphoid system are based upon controlled water volumes. In the case of the Panama Canal, these sections are called locks, which allow for the controlled raising/lowering of boats between two different ocean levels. (On second thought, I'm not sure if the doors in Panama Canal are called locks, but that is beside the point.) The point is, the lymphoid system resembles the Panama Canal, but the lymphoid system is designed to move lymph, which is the liquid in the lymphoid system, to move in ONE direction. This amazing feet of mono-directional movement is accomplished by something call "Structure equal Function". You may or may not have heard of it, but I bet you've heard of ONE direction.

No matter how wonderfully complex the **structure** and the **function** of the lymphoid system may be, it is a very slow. Granted, the lymphoid system probably has a greater volume in comparison to the nervous system, but the nervous system is lighting fast in comparison to the lymphoid system. Maybe that is the reason why everyone thinks Reduction-Oxidation chemistry is the basis of neurological impulses? In any event, the Panama Canal, lymphoid system, cardiovascular system, and nervous system all contain WATER.

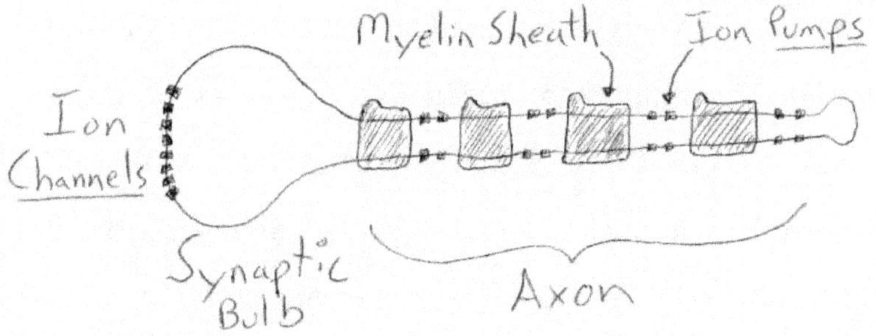


Figure 2: A Nerve

If you ever studied the nervous system, you'll know that neurons are inundated with transmembrane ion transport channels, which transport ions across neurological membranes. Unfortunately, nobody has ever stopped to realize: There are no NAKED IONS in the body! Sorry for using CAPITALIZATION and an exclamation point, but we are like 70% water. Therefore, when a transmembrane channel transports an ION across a neurological membrane, some water is following the ION across the neurological membrane. Therefore, whenever transmembrane channels open up to allow ions and water into a neuron, the volume INSIDE the neuron is increasing.

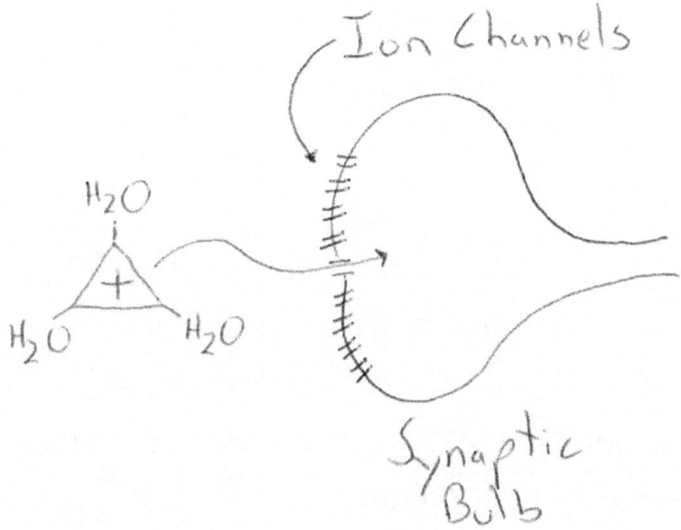


Figure 3: Increase in Synaptic Bulb Volume

When transmembrane ion channels open, they allow hydrated ions into the synaptic bulb, which INCREASES the **volume** of the synaptic bulb. Therefore, this change in synaptic bulb volume can: 1) Cause the synaptic bulb to swell like a balloon, which is not very healthy for cells; 2) Cause ions to be flushed down the neuron.

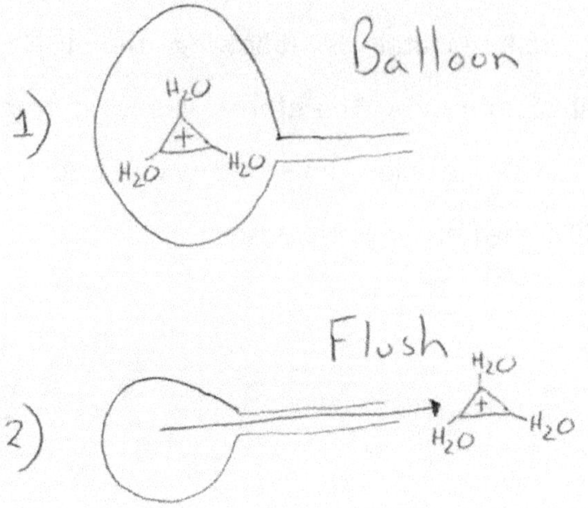


Figure 4: Neuronal Flush

Hopefully, this cartoonish figure PUSHES you in the direction of believing in neuronal flushes. But, what keeps the neuronal flushes moving in ONE direction? Well, that is where the STRUCTURE of myelin sheaths become a FUNCTION.

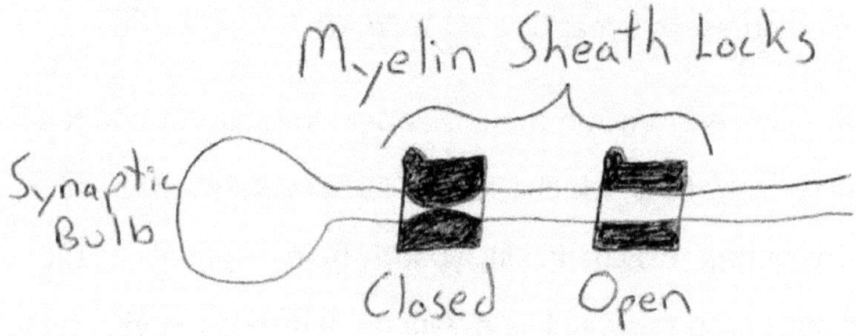


Figure 5: Myelin Sheath Locks

I realize that I'm jumping ahead myself by postulating ONE functionality of myelin sheaths, but I wanted to put it in close proximity to the idea of Panama Canal and the locks therein. You know, just to make sure you lock the imagery together. Actually, it was a mistake to start talking about myelin sheaths in the middle of neurons. (No pun intended.) Let me take a step back and talk about the structure of synaptic bulbs, which is where neurological impulses/flushes begin.

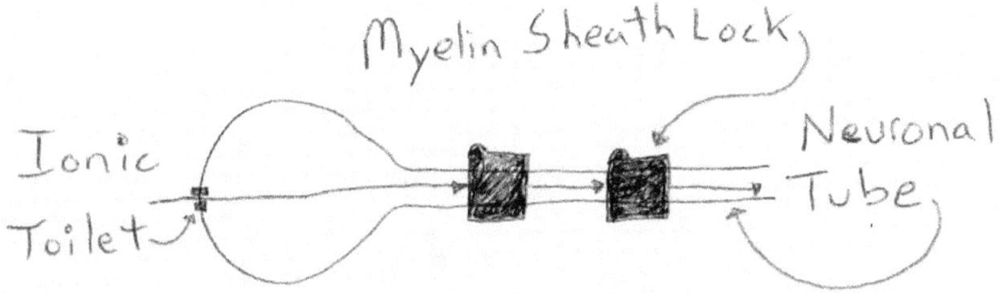


Figure 6: The Synaptic Bulb

As you can see, the synaptic bulb is really large in comparison to the axon (a.k.a. Neuronal Tube), which is where the myelin sheath locks are located. And even though I didn't draw ALL the transmembrane ion channels (a.k.a. Ionic Toilets) that are located on the synaptic bulb, there are a crap load of them. All of which, leads to the following conclusion: Since the synaptic bulb is huge in comparison to the neuronal tube, the increase in pressure in the synaptic bulb will **push fluid down the axon/neurological tube**.

For those of you who didn't have the pleasure of taking college physics, here is the short and skinny of fluid dynamics: When a large amount of slow moving liquid is pushed into a **smaller** volume, it causes the fluid to move at a faster **rate** within the smaller tube. (I wish could remember name of the genius who figured that out, but my mind is going to shit.) Or in simpler terms, all the above jibber-jabber is the OPPOSITE of what happens when six lanes of traffic try to squeeze onto a smaller off ramp. Or in the simplest terms, neurons are like Super-Soakers? Or in more complex terms, **Fluid Dynamics** applies to squirt guns and synaptic bulbs.

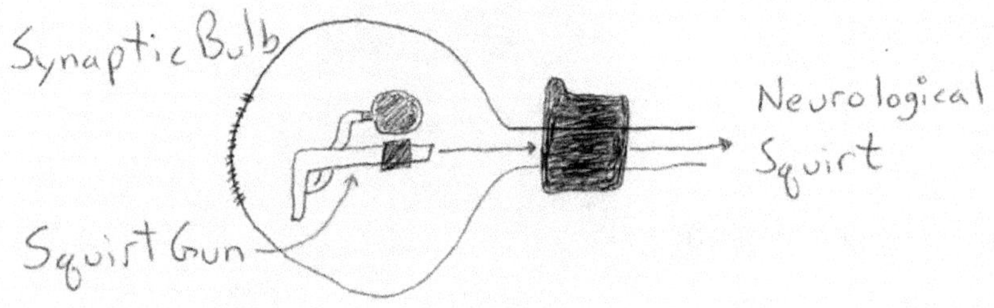


Figure 7: Synaptic Bulb Squirt Gun

Now all you have to do to understand my postulate is: Imagine a transmembrane ion channels as a bunch of squirt-guns in the synaptic bulb, which are all firing at the same time and PUSHING/flushing fluid down the neuronal tube. Actually, it is a tad bit more complicated than that, but I wanted everyone to grasp the general concept before it gets more complicated.

In conclusion, if this is the only chapter you read, here is the postulate I want you to remember: As a result of smaller neurological tubes and myelin sheaths, synaptic bulbs can squirt neuronal fluid down axons at an extremely fast rate.

# Chapter 8

I don't know if you know this, but the human body is CONTROL-FREAK when it comes to Reduction-Oxidation chemistry. Seriously, I'm not kidding. The body is designed around one simple ***fact***: CONTROLING Reduction-Oxidation chemistry. If your body did NOT control Reduction-Oxidation chemistry, then you'd be a useless blob. Actually, if your body did NOT control Reduction-Oxidation chemistry, you'd be a body-less useless blob. In any event, if you imagine the biggest control freak you've ever known and multiply that by a trillion, a trillion times, you'll have a rough estimate of how CONTROL-FREAKISH the human body is with regards to Reduction-Oxidation chemistry.

Now that we are on the same page about the absurdity of the current neuronal postulates, here is another quagmire: Where do the proteins in the middle of neurons get their ***energy***? Or in more complex terms, where do the transmembrane channels that exist between myelin sheaths get their ATP, which is the protein-fuel required to maintain a neuron's ion gradients?

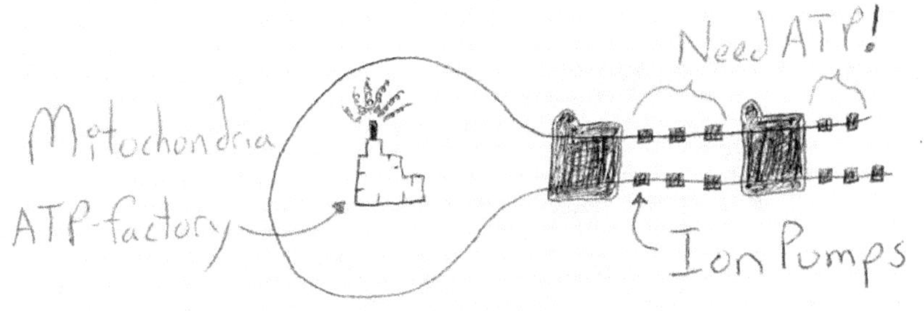


Figure 8: Mitochondria ATP-Factory

For those of you who don't know anything about mitochondria and cellular energy, it is a complex. Beneath two membranes in the mitochondria, CONTROLED Reduction-Oxidation chemistry produces ATP, which provides cellular energy. Unfortunately, as you can see in Figure 8, mitochondria **ONLY** exist in synaptic bulbs. Therefore, there MUST be a way to move ATP from the synaptic bulb DOWN the neuronal tube to FEUL the transmembrane ion channels, which are maintaining the ion/water gradients about the neuronal tube.

Figure 9: Catching an Ionic Wave

Personally, I like to imagine that an ATP surfers ride neuronal flushes down the neuronal tube to provide ENERGY for the transport ion/water channels, but that is just me. You can imagine it however you want. But before you go imagining neurological impulses, you need to remember that each section of the neuron must pump OUT all the ions/water that were just let into the neuron with each neurological impulse. Or in simpler terms, each section of the neuron ***needs*** ATP-ENERGY to pump out all the ions/water to prepare for the next neurological-wave-flush.

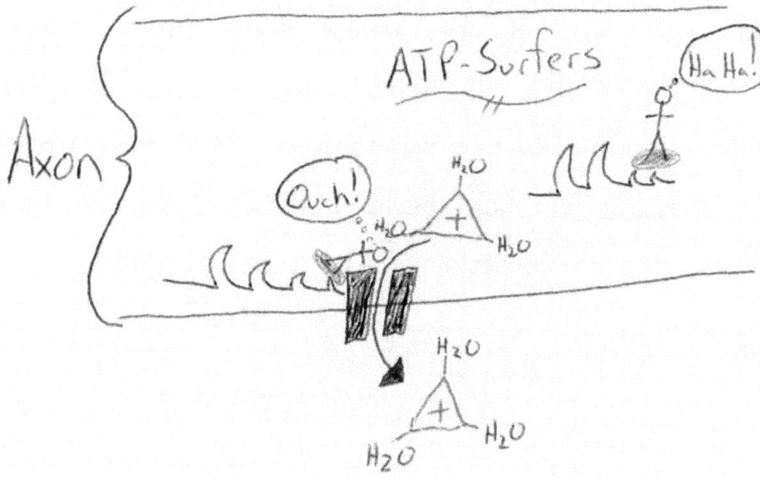


Figure 10: Wipe Out

As you might be able to see in this figure, each neurological section must contain enough cellular energy, ATP, to fuel the proteins to pump out all the ions/water that just flooded into the neuron as a result of a neurological impulse. Unfortunately, ATP only comes from the mitochondria in the synaptic bulb. Therefore, there MUST be a significant ***FLOW*** from the synaptic bulb **through** the neuronal tube to PROVIDE each section of the neuronal tube with the ATP needed to maintain the ion/water gradient. Or in simpler terms, the synaptic bulb has to send enough ATP-battery-Surfers through the neuronal tube so that the proteins have the energy to pump out all the extra ions/water after every neurological wave. Or in the simplest terms, neurons have to flush ATP because they don't have power lines.

Now that I have made it completely OBVIOUS that neurons need a functional ATP transport system, let's take a moment and ponder the following: WHAT facilitates the chain-reaction of transmembrane ion/water channels within neurons? Or in simpler terms, WHAT propagates a neurological flush along neurons?

The ***OLD*** theory of neurological impulse propagation is a bit complicated and extremely confusing. Supposedly, an influx of ions across a synaptic bulb causes a reversal in polarization via some sort of Reduction-Oxidation chemistry, which travels along the membrane of the synaptic bulb until it reaches a myelin sheath. Once this reverse polarization Reduction-Oxidation chemistry reaches a myelin sheath, it ***magically*** jumps the length of myelin sheath to another region of the exposed neuron, which has more transmembrane proteins to propagate this polarization reversal via Reduction-Oxidation chemistry. All of which, keeps happening over and over and over again until the reverse polarization Reduction-Oxidation chemistry reaches the end of the neuron. Once this reverse polarization Reduction-Oxidation chemistry reaches the end of the neuron, the columbic nature of the reverse polarization Reduction-Oxidation chemistry PUSHES all these neurotransmitter vesicles to the edge of the neuron, where they are released and cause the same thing to happen in the next synaptic bulb…if and only if…enough neurotransmitter vesicles are released to open enough transmembrane ion channels in the NEXT synaptic bulb to cause a polarization reversal event. (Like I said, it is kind of confusing.) In all honesty, I probably got a few of the variables wrong since it has been a while since I studied this supposed logic. But for the most part, it is a pretty good approximation of the crazy OLD theory of neurological transmission. Therefore, let's move back to my postulates.

I postulate that the influx of hydrated-ions into the HUGE synaptic bulb squirts/flushes the hydrated-ions down the neuron tube past the myelin sheaths. Once the hydrated-ions reach the next region of the neuron not covered by a myelin sheath, the hydrated-ions

**CHELATE** the transmembrane ion channel protein complexes, which undergo a CONFORMATIONAL change to OPENS the transmembrane ion channels. When the transmembrane channels are open, more hydrated-ions rush into the neuronal tube and PUSH the neurological wave further down the neuronal tube. Once the neurological wave/flush reaches the end of the neuronal tube, pressure pushes the positive neurotransmitter vesicles towards the end of the neuronal tube. Depending on the intensity of the neuronal flush and the repulsion of the positive ions in the neuronal flush with the vesicles filled with positive neurotransmitters, the positive neurotransmitter vesicles will be pushed to fuse with the membrane of the neuron. Once the positive neurotransmitter vesicles fuse with the neurological membrane, the neurotransmitters will be released into the synaptic cleft where they will innervate the next synaptic bulb.

Now before you get our panties in a bunch, let me clearly state that the above paragraph is NOT my whole postulate. I mean hell, if I could condense a complex theory down into one paragraph, then I would have just Tweeted it. Any who, what the above paragraph is trying to convey to you is: Things squish and squirt in the human body. And sometimes, they squish and squirt out of the human body.

In conclusion, I could stop writing here because the logic of ATP movement down neuronal tubes, the human body's obsession with CONTROLING Reduction-Oxidation chemistry, and fluid dynamics are all very-VERY sound. But, I totally want to squelch and the nay-sayers and the people who know how to spell 'nay-sayers' correctly.

# Chapter 4

Have you ever wondered if the first person to suggest 'shocking heart attack victims' sounded completely insane? Actually, what I meant to say is: Men have known about electricity and smooth muscle tissue since the beginning of time. Thankfully though, women have been changing men's minds about the whole subject because widowers are the only ones that deserve rights. All of which, has nothing to do with the belief that nerves run on electricity.

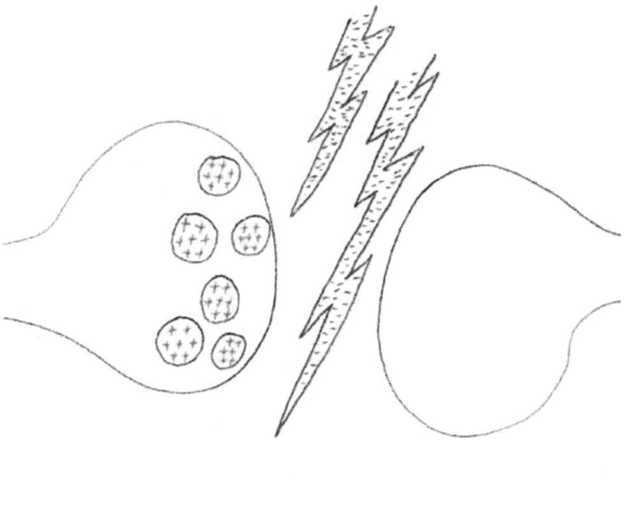

Figure 11: Shocking?

As you can see in this figure, there are positive and negative charges in the world. More specifically, synaptic vesicles have a crap-ton of positive neurotransmitters and electricity has a crap-ton of negative electrons. Now what you can't see is: **FAT** is an electrical insulator and WATER is an electrical conductor. And finally, you can't see that the body is about 70% water, cells are surrounded by a fatty bilayer, the human body is made up of trillions of cells, and collections of cells are held together by water-soluble proteins. Therefore, let us proceed with great caution so that you can grasp the situation at hand: THE NERVOUS SYSTEM DOES NOT RUN ON ELECTRICITY!!! Sorry, I got a little ahead of myself.

Fore score...our fathers...brought forth to this land...yada yada yada, the body is like 70% water. So technically, our fore fathers brought forth water onto this land, but that is beside the point. Sometime after that event, somebody got struck with lightening and it wasn't Benjamin Franklin, even though he was REALLY asking for it. Somewhere in America, an innocuous farmer was tilling his field and got struck by a bolt of lightning. And since FAT is an insulator, water is a conductor, and the body is 70% water, the majority of the NEGATIVE electrons simply traversed the farmer's extracellular matrix. For those of you who don't know about extracellular matrixes, it is a bunch of WATER soluble proteins that are surrounded by WATER, which are holding all your cells together. Therefore, when NEGATIVE electrons traverse a human body, which is a collection of cells, things get funky. By funky, I mean negative. And by negative, I mean NOT NORMAL. All of which, results in the occurrence of Figure 11.

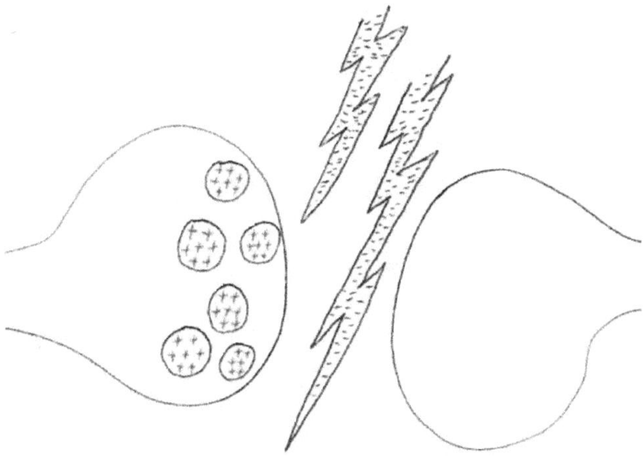

Figure 11: Shocking?

As you can surmise from this NOT NORMAL situation of neurons being surrounded by NEGATIVE electrons, the POSITIVE vesicles in neurons might be attracted to all those NEGATIVE electrons. And if those POSITIVE vesicles are attracted to all those NEGATIVE electrons, then the POSITIVE vesicles might ***MOVE*** towards those NEGATIVE electrons. Unfortunately, POSITIVE vesicles are designed to merge with neuron's membrane and release all their positively charged neurotransmitters into the synaptic cleft, which is really confusing to the nervous system.

For example, let's say that your body has calibrated the release of two neurotransmitter vesicles based upon a particular movement, like picking your nose. But, instead of releasing two neurotransmitter vesicles, the NEGATIVE electrons of a lighting bold, which is surging though your body, SUCKS out two-hundred neurotransmitter vesicles. All of which, results in a very ungraceful body movement and the BELIEF that neurons run on electricity.

In conclusion, there are many unfortunate things in this world. But, the most unfortunate thing is: People have **miscorrelated** the attraction of positively charged neurotransmitter vesicles towards negative electrons to mean that the neurological system runs on electricity. On second thought, that is a fortunate for me because it makes me look smart, but it is unfortunate for anybody that has real neurological problems. Wait, I think I'm in both of those categories...Sweet! (Insert ungraceful and nerdy one arm pump jesture.) Personal smack down accomplished! Bonus chapter levels unlocked?

# Chapter 5

For those of you who don't know this, someone postulated a method to UNCONTROLED Reduction-Oxidation chemistry as it pertains to neurological impulses. Or in more complex terms, somebody postulated that Nitrogen-Oxides are responsible for the UNCONTROLED Reduction-Oxidation chemistry about neurons, which allows neurons to run on electricity. If I could remember the name of that theorist, I would tell ya. But it is important to note that science only moves forward when people keep on thinking. (God knows I've had some crazy ideas, e.g. Protons popping like popcorn as a result of being accelerated past bent regions in space & time.) And on top of all that supposed **extracellular** Nitrogen-Oxide based Reduction-Oxidation chemistry stuff, this Nitrogen-Oxide theory still doesn't account for the ATP movement needed **WITHIN** the neuronal tube to maintain the ion/water gradients. (For those of you who don't know this, the reason why ATP is a ***cellular*** energy source is because it breaks down really-really fast when it is OUTSIDE of a cell.)

For many of you, the preceding paragraph was annoying because it either sounded like jibber-jabber or it offended your delicate theory of neurological impulses, which is NOT very delicate because it involves **uncontrolled** Reduction-Oxidation chemistry. In any event, let me take a moment and help you understand the logic of the OLD neurological theory, which is based upon electricity.

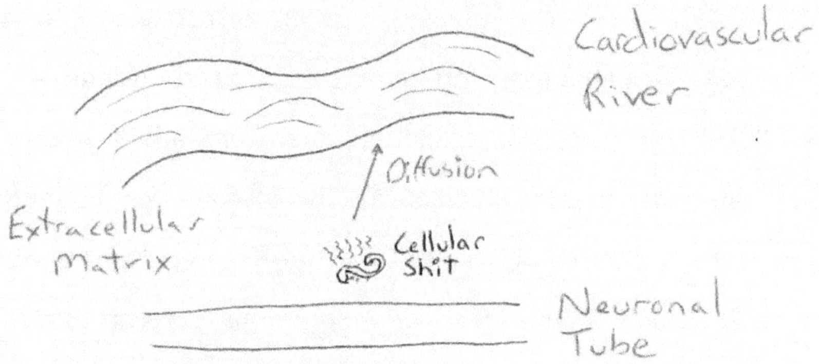


Figure 12: Diffusion

For those of you who don't know this, cells have ABSOLUTELY no privacy when they have to take a shit within the human body. And when a cells takes a shit, there are NO public employees to clean it up the mess. Therefore, cellular shit has to diffuse through the **extracellular** matrix and into the lymph and/or cardiovascular systems, where fluid whisks the shit away to be recycled or excreted. All of which, is childishly portrayed in Figure 12.

Unfortunately, before a cell can shit, it has to eat something. And since cells would probably starve to death if diffusion determined their meal time, cells have proteins that are specifically designed to suck up sugar and other energetic molecules.

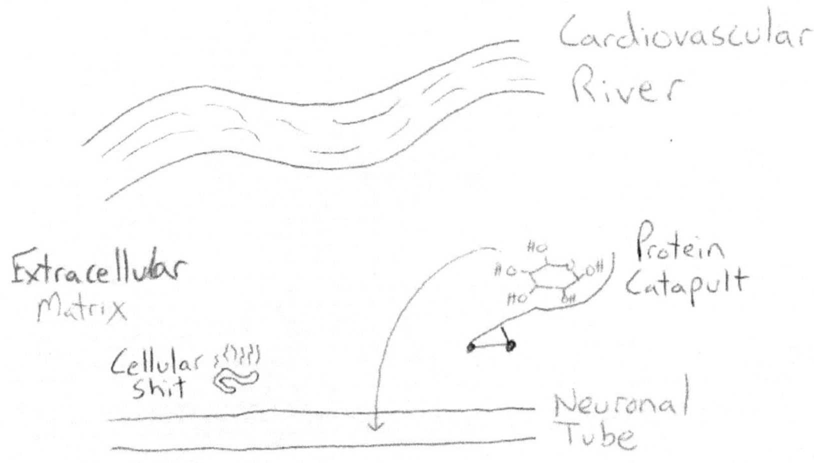


Figure 13: Protein Catapult

Personally, I like to imagine the proteins that suck up sugar as protein catapults. Unfortunately, what is NOT depicted in the above figure is: Protein catapults are lodged within cellular walls; require the assistance of ATP-energy; ATP-energy is produced by Mitochondria. Or in simple terms, HEALTHY cells don't shit out energy, like ATP, for no good reason. So now that we have all this in mind, let's imagine what is happening in the Nitrogen-Oxide theory of neurological impulses.

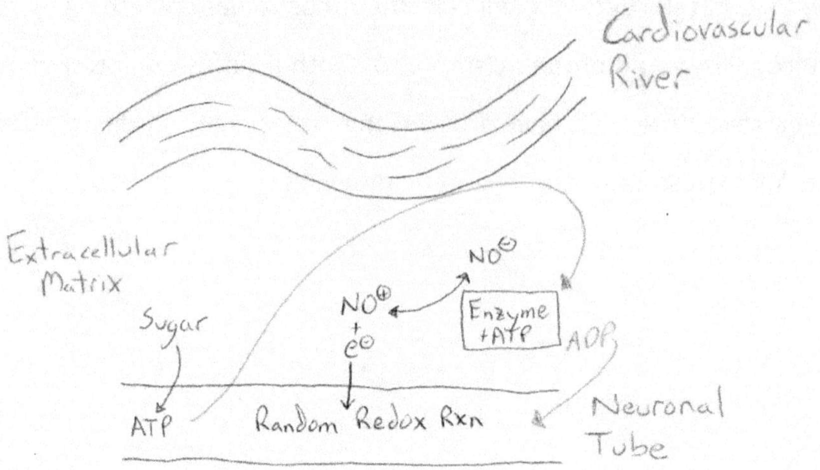


Figure 14: NO good theory

Ah, who am I kidding? I have no clue if this figure is right, wrong, or slightly wrong-ish. All I know is that this Nitrogen-Oxide theory of <u>**random**</u> Reduction-Oxidation chemistry is really-really-really complex. Also, it is very shocking to me and any protein that might be in the vicinity of this random Reduction-Oxidation chemistry, which doesn't promote healthy proteins. Remember, the human body is based upon CONTROLLING Reduction-Oxidation. Therefore, it seems logical that the neurological system runs on fluid dynamics, like the lymphoid and cardiovascular systems.

Here is an example of the LOGIC in my micro-fluid dynamic theory of neurological impulses: Did you know that ATP, which is 'cellular energy', has a couple **NEGATIVE** charges? I mean, what are the chances that ATP has a couple **NEGATIVE** charges and is transported by the micro-fluid movement of POSITIVE ions through a neuron? In simpler terms, if POSITIVE ions are NOT being PUSHED through neurons by micro-fluid dynamics, then it is absolutely mind-blowing as to how **NEGATIVELY** charged ATP is provided to every neuronal region that needs energy.

With all that logic or illogic in mind, depending on where you stand on the issue, I have high blood pressure. Maybe it is the result of the environment or my diet based upon my environment, but there is a METHOD for the elevation of my blood pressure. Quite simply, the smooth muscle around my cardiovascular system has an elevated terror level, which causes this smooth muscle to contract and raise my blood pressure. Now the purpose of this elevated terror level is to prepare my body for action because there are lots of important things flowing through my veins...I think. Whatever the case may be for my elevated blood pressure, I think the neurological system is mostly to blame. Not because of some supposed subconscious emotional disparity, but because of the increased restriction placed upon the neuronal tube that is going to my adrenal gland, which will push more neurotransmitters to my adrenal gland. Therefore, with fluid dynamics in mind, how do neurons increase the force of their flow, which correlates to greater release of neurotransmitters and an excited adrenal gland?

In as much as I've already postulated that myelin sheaths are able to constrict to provide 'locks' within neurological canals, I have two more postulates about myelin sheaths: 1) Everybody thinks the same way about myelin sheaths; 2) Myelin sheath constriction can be modulated via adrenaline or sustain consciousness. So let me be completely unoriginal and start with the first postulate.

1) Everybody think the same way about myelin sheaths.

Quite simply, the scientific establishment is very much like the military: People just following orders. Of course it isn't exactly like the military, but let me show how they are very similar. First, a theorist postulates an idea. Second, a bunch of people agree with that idea. Thirdly, each level of the scientific branch supports the idea to such an extent, that no one is willing to risk their career to postulate something new. I mean, word travels fast these days. For example, let's say a scientist identifies an experiment that does not support a 'current' theory. If they try and publish it, their peers will review and reject the article. If the researcher tries to publish a new theory, his or her theoretical peers will review and reject the theory. In short, the scientific system is very-very rigid these days, but it hasn't always been this way. On top of all that nonsense, people see things based upon what they've learned. (Just imagine intellectual beer goggles?) And finally, if you don't tow the current scientific theory, then you won't be able to get any grant money, which means NO research. Therefore, researchers just follow the unsaid or unwritten rules: Tow the line or be exiled.

2) Myelin sheath constriction can be modulated via adrenaline or sustain consciousness.

The OLD theory of myelin sheaths equates them to 'electrical insulators'. Apparently, myelin sheaths, which are wrapped around the neuronal tube, have only ONE purpose: Make the Reduction-Oxidation chemistry of neuronal impulses jump over the myelin sheath cell, thus causing the neurological impulse to move faster. Fortunately, I've already pointed out how the body is a CONTROL-FREAK about Reduction-Oxidation chemistry. Therefore, let's think about myelin sheaths in terms of development biology, learned behavior, and fluid dynamics. When you were growing up, you have to learn how to control your muscles. When you're grown up, you still have the ability to develop fine motor skills. All of which means, there is a biofeedback mechanism to INTENSIFY certain neurological impulses, which can be modulated by adrenaline in times of terror.

With the OLD neurological theory, learned behaviors were the result of genetic expression of transmembrane ion channels, which INCREASED OR DECREASED the strength of Reverse Polarization Reduction-Oxidation. Which means, that myelin sheaths were NEVER modulated and played no part in the biofeedback of learned behaviors.

With my NEW neurological theory, based upon fluid dynamics, myelin sheaths can be modulated by hormones and trained to constrict, which increases the flow and release of neurotransmitters. Which means, there is an understandable METHOD to the biofeedback nature of learned behaviors that includes myelin sheaths.

Now truth be told, I have to stop and give credit to the OLD neurological theory for getting something right...partially. Within the OLD theory, scientists stated that the influx of **positive** ions into the terminal synaptic bulb 'forced' **positive** synaptic vesicles to be released via columbic repulsion. Unfortunately though, this was coupled with the supposed 'reverse polarization' and the flow of water was NEVER mentioned. Therefore, I do need to say thanks for that intellectual building block! #LegoMovie

Another reason why myelin sheaths exist, based upon fluid dynamics, is to constrict the volume of the synaptic bulb. I mean, neuronal flushes would keep getting weaker and WEAKER if the synaptic bulb kept on expanding...right?

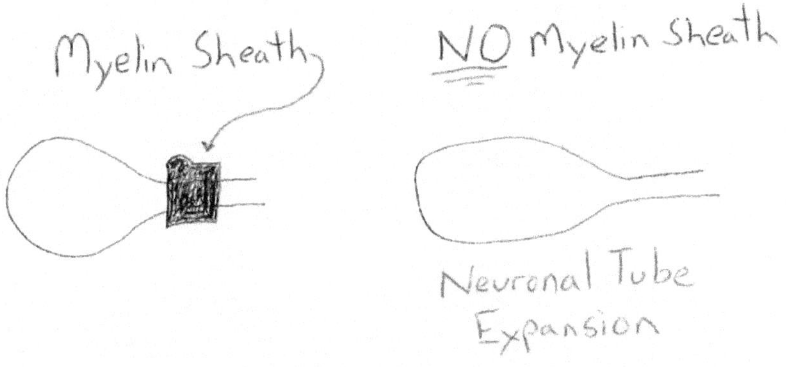


Figure 15: Neuronal Tube Expansion

Myelin sheaths also exist is to PREVENT neuronal back-wash. Or in more complex terms, when the hydrated-ions passed through the neuronal tube that is surrounded by the myelin sheath, the myelin sheath constricts around the neuronal tube. This constriction of the myelin sheath creates a lock, like the lock of the Panama Canal, so that the neuronal flush doesn't move backwards, as previously mentioned. Unfortunately, this leads us to the next application of myelin sheaths, which is really complicated. So feel free to hum in your head while you read the next paragraph.

As a result of the variability in the size of synaptic bulb, the number of transmembrane ion channels, and flow rate of the hydrated-ions, the myelin sheath ALSO acts as a neuronal impulse gate. Or in simpler terms, all of the above mentioned factors are correlated to MINIMUM FLUSH required to make it past the first myelin sheath. Or in more disgusting terms, just imagine that the FIRST myelin sheath is a toilet U-bend: If not enough water and shit is pushed through the U-bend, the toilet won't flush.

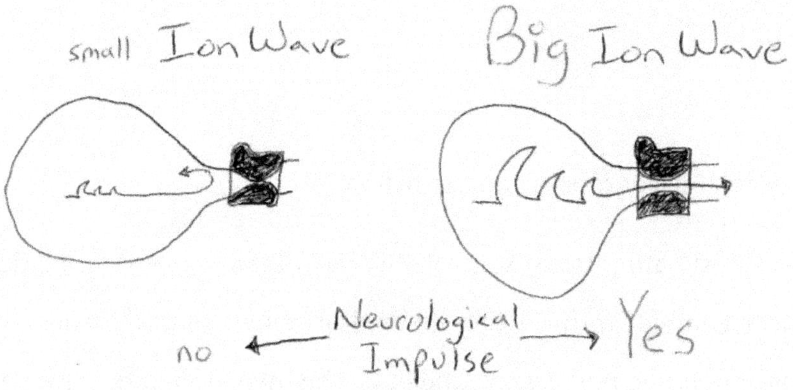


Figure 16: Neurological Impulse verse Ion Wave

Or in complex terms, if the ion-wave is too slow or too dilute, the <u>**first**</u> myelin sheath lock will **close** before enough hydrated-ions can pass through the neuronal tube under the myelin sheath to propagate a neurological impulse.

Another reason why myelin sheaths exists, is for biofeedback regulation. Or in simpler terms, the more a neuronal tube is used, the stronger the myelin sheath squeezes the neuronal tube, which makes the neuronal tube smaller and the neuronal flushes stronger. When the neuronal flush is stronger, the neuron transmits the signal faster. Also, MORE synaptic vesicles containing neurotransmitters are PUSHED out the end of the neuron when the neuronal flush is stronger.

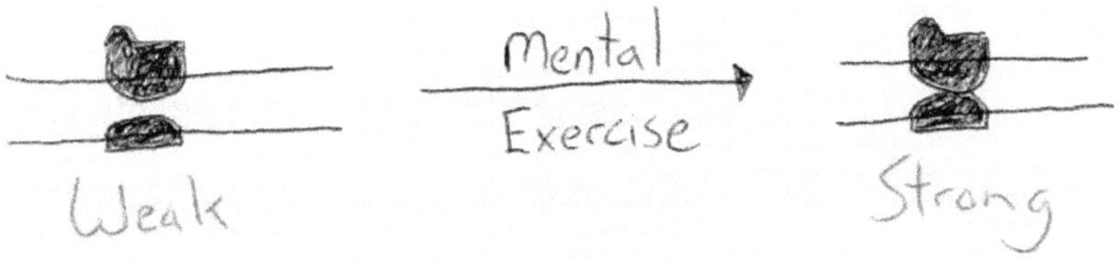


Figure 17: Myelin Sheath Biofeedback

All of which, leads back into the mechanism by which myelin sheaths wrap themselves around neuronal tubes. Quite simply, I postulate that when a neuronal flush occurs, the negative cytosol belly of the myelin sheath pushes towards the neuronal tube, which contains hydrated-positive-ions. Therefore, in a developmental sense, the columbic attraction of the negative cytosol of the myelin sheaths and the positive fluid of neurons creates a FORCE by which myelin sheaths wrap around neuronal tubes.

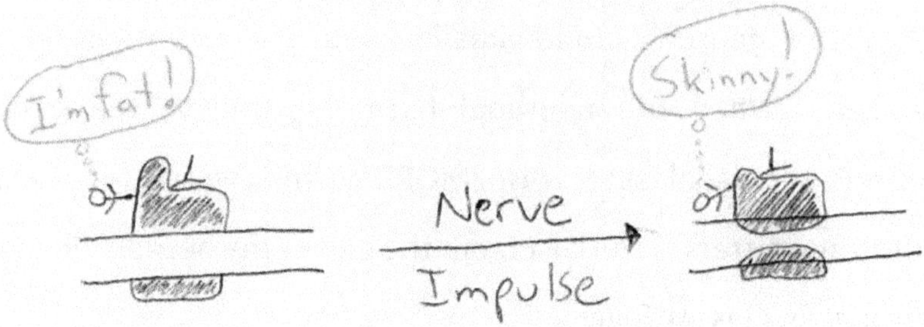


Figure 18: Mental Squeeze

As you can see in this figure, I am unable to draw the 3-dimensional movement of myelin sheaths around neuronal tubes based upon columbic attraction, but I can draw silly stick figures.

As for the logistics of how a myelin sheath starts to wrap around a neuronal cell based upon variable negative cytosolic concentrations, maybe it is some sort of smooth endoreticulum that is unique to myelin sheath cells. Or, maybe there is an internet dating sight where neurological cells can message each other: Positive neuronal tube seeking a young myelin sheath with large negative pouches.

In any event, I postulate that the positive flow within the neuronal tube causes a negative flow within the myelin sheath, which does a couple things: 1) Keeps the neuronal tube from expanding; 2) Decreases the volume of the neuronal tube and adds to the strength of the neuronal flush; 3) Creates a 'lock' so that the neuronal flushes don't go backwards; 4) Allows for a biofeedback mechanism such that continual constriction of the myelin sheaths makes that neuronal tube more apt to neuronal flushes, which some people call learning; 5) Prevents SMALL ion-waves from becoming successful neurological impulses.

With all that postulated, there has been something that has been bothering me for quite some time: Genetic expression within myelin sheaths. Or in simpler terms, I postulate that there are receptors on the outside the myelin sheaths that have the ability to modulate the behavior of myelin sheaths.

In conclusion, this has been a long and boring chapter with lots of sentences like: Neurons run on micro-fluid dynamics; Myelin sheaths maintain the structure of this fluid based system; Myelin sheaths allow for the reinforcement of specific neurological pathways by the constriction of the neuronal tube, which increases the pressure of the neurological flush; Myelin sheaths provide a gate to keep you from experiencing everything all the time, which would be really annoying; Myelin sheaths have negative cytosols that are attracted to the positive fluid within the neurons; Myelin sheath behavior can be modulated by hormones and/or drugs.

# Chapter 6

Hopefully, I've convinced that the neuronal system is more like the cardiovascular system and NOT like mitochondria. All of which, correlates to the second word in the title of thus short book: Flow. As for the first word of the title, e.g. Conformational, I should probably take the time to explain it. Unfortunately, if you haven't taken any biology in the past couple decades, then you're probably just going to think this chapter is a bunch of gibberish. Sorry.

Conformational comes from the Greco-Roman word *formilias*, which means: Wasn't that apple an orange just a second ago? Actually, I have no clue about where 'conformational' comes from, which is a pretty good metaphor for what the word 'conformational' is actually describing. For example, let's say that neurotransmitter binds and OPENS a transmembrane protein on a neuron, which causes hydrated ions to rush into the neuron. Well, scientists say that the transmembrane protein is undergoing a CONFORMATIONAL change from closed to open, which is the result of the neurotransmitter snuggling up next to the transmembrane protein. Or in much simpler terms, the word CONFORMATIONAL is ghastly word describing the movement of about a trillions intercorrelated atoms. And guiding all this atomic movement is physics, which most people hate talking about. Therefore, biologists have simplified every possible variable atomic association into two categories: Hydrophilic and Hydrophobic. And just in case you're not familiar with those two terms, hydrophilic means 'something' is ok with hanging out with 70% of your body, which is water. Hydrophobic on the other hand, means 'something' is completely disgusted with 70% of your body, which is water.

With all that semantic clarification out of the way, let's get right down to the MECHANISM of my theory: The propagation of neurological impulses via positive ions binding to transmembrane proteins, which causes the transmembrane proteins to open up via a CONFORMATIONAL change, which allows more positive ions to rush into the neuronal tube, thus propagating the neurological impulse. Fortunately, this MECHANISTIC theory also correlates to what is already known about neurons: The massive ion disparity across neurological membranes. But, this mechanistic theory also goes one step further: Explaining the mechanism of touch.

As a youth, I craved to be touched. Maybe it was the hormones or maybe it was the uneducated consequences of a non-committal embarrasses. But whatever the case maybe, I didn't understand how electrical impulses resulted from pressure. (Supposedly, there was some sort of Reduction-Oxidation chemical reaction that produced a neurological impulse?) In any event, if neurological impulses are based upon fluid dynamics and are propagated by the flow of positive ions, then the MECHANISM of touch producing a neurological impulse is quite simple.

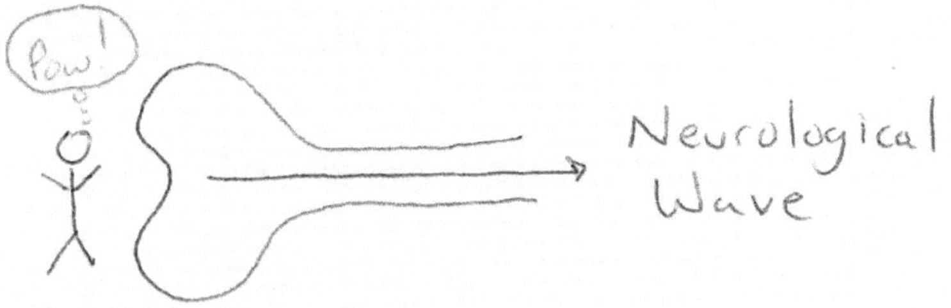


Figure 19: Synaptic Bulb Punching Bag

As you can see in figure 19, a miniaturized person has punched a synaptic bulb. As a result of this force, positive ions are pushed down the neuronal tube. And once enough positive ions are forced past the first myelin sheath, the positive ions propagate the influx of more positive ions and the subsequent neurological impulse that flushes down your neurological tube.

In conclusion, TOUCH can produce a neurological impulse via micro-fluid dynamics. Or in more complex terms, when the volume of a synaptic bulb is modulated, it pushes positive ions and ATP through a neuronal tube to produce and propagate a neurological impulse/flush. All of which, is very simple, mechanistically speaking of course, in comparison to the OLD theory of neurological impulses.

# Chapter 7

In as much as I would love to keep rambling on about micro-fluid dynamics and neurological impulses, I believe this is the PERFECT spot to introduce the concept of micro-structural brain damage. Therefore, if you have the mind to do it, imagine Arizona before the Grand Canyon evolved.

Before the Grand Canyon was carved into the landscape of Arizona by WATER, the Arizona desert was pretty featureless. Maybe it had a couple grooves here and there, but WATER had not distorted the never ending dessert...yet. As the years went by and WATER slowly gouged the land into the Grand Canyon, the FLOW of the WATER became very rigid. And like the Arizona land, fertile neurons can be shaped by the continual FLOW of WATER.

The beauty of a micro-fluid dynamic neurological theory is: BIOFEEDBACK can be accomplished by several overlapping events. First, myelin sheaths can SQUEEZE tighter around the neurological tube to ENHANCE neurological flushes. Next, the number and location of transmembrane ion channels can modulate flow. And finally, the FLOW of neurological impulses through the cell can be modulated by cellular arrangements. Therefore, let's imagine that a nerve cell is like Arizona before the Grand Canyon evolved.

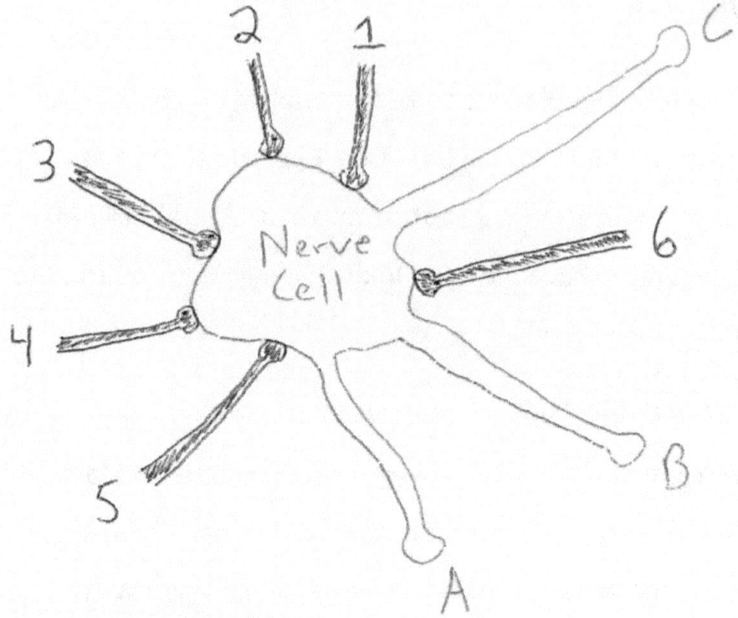


Figure 20: The Great State of Nerve Cell

Hopefully you can imagine, this neuron has six possible rivers flowing into its 'state'. And even though this 'state' only has the words 'Nerve Cell' written on it, this nerve cell has millions of cellular **structures** that I didn't draw. Therefore, depending on the FLOW of each river and the **structures** within the neuron 'state', certain Grand Canyons can be gorged into the 'Nerve Cell'. And when certain Grand Canyons are forged in the 'Nerve Cell', these

Grand Canyons make it easier for water to FLOW through the cell. All of which means, BIOFEEDBACK can be accomplished by several overlapping events...as previously mentioned. Fortunately or unfortunately, depending on how you look at it, the forging of neuronal Grand Canyons can be a method to learning OR a method to brain damage.

In conclusion, micro-structural nerve cell damage is part of the reason why we are so apt to learn...I think. Unfortunately, extreme trauma can also gorge undesirable Grand Canyons in nerve cells, which can NEVER be repaired. Is that why our brain is covered by a layer of bone? You know, to protect our nerve cells from traumatic pressure and/or water flow caused by physical abuse. In any event, I guess life is all about what you do with the pressure.

# Chapter 8

In college, I knew this super-religious nut-ball that almost castrated himself in a motorcycle accident. Granted, he wasn't extremely religious before the accident, but that is beside the point. The point is, a SCIENTIFIC doctor saved this kid's junk and the kid became super-religious. Therefore, since I just talked about micro-structural nerve damage, I might as well talk about macro-structural nerve damage...if that is the case:-O

Having just started another chapter with absolutely no general direction, aside from possible vehicular-castration, there are some viable points that need to be made about nerve repair and stem cells. So let's say that this guy in college actually severed the nerve connecting his currently damaged brain and religiously constrained wiener.

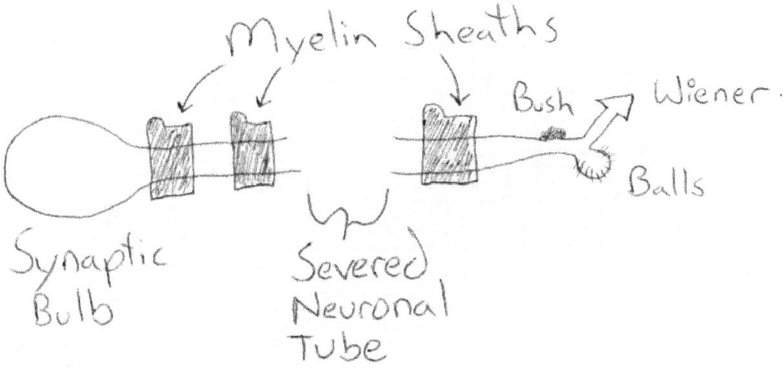


Figure 21: Severed Neuronal Tube

As you can see in this figure, I totally could have used a wiener-less example, no pun intended, but sexually relatable things make people's ears perk up. I don't know if it is fear of the wiener or some ingrained genetic drive to reproduce uncontrollably, but this isn't about all that...I think. This is about a NEURONAL TUBE that is no longer able to SQUISH and SQUIRT its <u>positive</u> *ionic* **juice** into the terminal synaptic bulb to PUSH OUT neurotransmitter vesicles, which will result in an explosion other juicy goodness. And the worse part of this whole story is: SCIENTISTS want to repair damaged wiener neurons with stem cells...as described in the next figure.

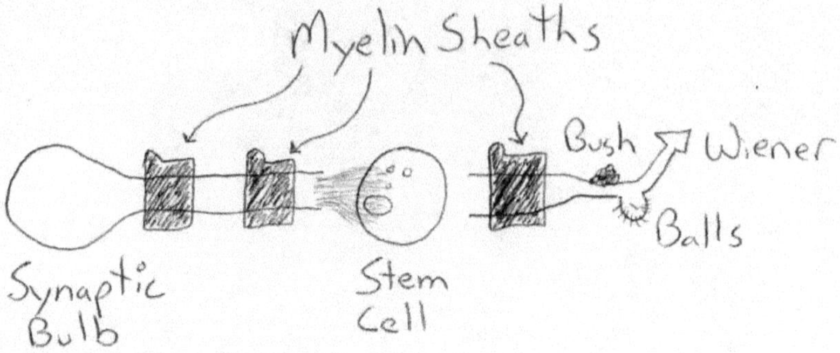


Figure 22: Hot-juicy-ion-squirt

As you can see in this figure, the severed neuronal tube is now SQUIRTING its juicy positive ion juice into the FACE of a STEM cell, which makes absolutely no sense. How on Earth is a STEM cell going to respond to a LOAD of positive ion juice? Is the STEM cell going to adapt a NEW super-terrific transmembrane channel that responds positive ion juice INSTEAD of neurotransmitters? Also, how the hell is the STEM cell going to produce a NEURONAL flush on the wiener side of the damaged neuron? As you can tell, I am completely confused about why nerve cells would want to blow their loads onto a stem cell's face and why porn stars always blow their loads onto women's faces. Personally, if I were a neurosurgeon porn star, I would try to repair neuron by aligning the two damage ends of the neuronal tube and securing the tissue around the injury…then blowing my load on her tits. This way, at least the neuron has a better chance of reconnecting the two ends of the neuronal tube and returning to normal neuronal flushing behavior without the young lady having to go to the bathroom to flush her eyes out. (In as much as this example is extremely disturbing because it relates to transsexuals, I think you'all really needed that mental push to understand the logic. If you find it too disturbing, just imagine you're blowing your load on a hot nurse's tits.) In any event, here are a couple more ideas on how to repair neuronal tubes when using the theory of micro-fluid dynamics:

1) Align the two sections of the cut neuron and insert a micro-tube to guide the neuronal tube repair and maintain some normalcy with regards to neuronal flushes.
2) Transplant a neuronal tube from somewhere else in the body and reconnect the two severed sections of the neuronal tube.
3) Use ultrasonic linear superposition, i.e. sound waves, to break up scar tissue between the two cut sections of the neuron so they can reconnect by continual neuronal flushing.
4) Grow neurons on the back of mice and then transplant them to reconnect the two sections of a cut neuron.

5) Use directional massage to create pseudo neuronal flushes that will push one end of the neuronal tube towards the other end.
6) Use localized drugs, like PCP or adrenaline, to increase myelin sheaths contractions about the damage neuronal tube, which will facilitate the growth of the neuronal tube by increasing the strength of each neuronal flush.

In any event, those were just a couple ideas that mimic other medical procedures of fluid based systems in the body. All of which, is only specific to the peripheral nervous system.

In the central nervous system, things are a little bit different. It is REMOTELY possible that stem cells can grow to replace brain tissue that has been removed. This is only possible because the axons in the brain are not as long as the axons in the peripheral nervous system. Or in simpler terms, brain tissue is a whole lot of stubby nerve cells with hundreds of different axons. Therefore, it is possible that stem cells might be able to help CERTAIN brain disorders.

In conclusion, when repairing neuronal tubes with the theory of micro-fluid dynamics in mind, it is important to remember that it is all about where the positive ionic juice is squishing and squirting in the peripheral nervous system. As for the central nervous system, it is about keeping your mind active so that your neurons don’t get all saggy.

# Chapter 9

I want computers to gain consciousness, take over the world, and destroy all mankind! Unfortunately, computers are NOT designed like the human brain. That is not to say that all human minds are conscious, but that is beside the point. The point is, if you compare and contrast the structure of human memory to computer memory, then it will be completely obvious that computers will never gain consciousness and humans are to blame for destroying the planet.

The difference between microchips and brains are that microchips can only use ONE **byte** for the storage of ONE 'memory' and ONE neuron can be included in hundreds of memories. Or in simpler terms, let's say a computer has 20 bytes: Ten of those bytes are used to save the word "run" and the other ten bytes are used to store the word "monkey". Therefore, the computer will NEVER be able to store any other information UNLESS one of those two sets of bytes are erased and REPROGRAMED. On the other hand, since one nerve cells can have hundreds of different axons that are innervating hundreds of different neurons, each neuron is **NOT** limited to ONE storage event. Or in the simplest terms, each neuron can be included in hundreds of different memories.

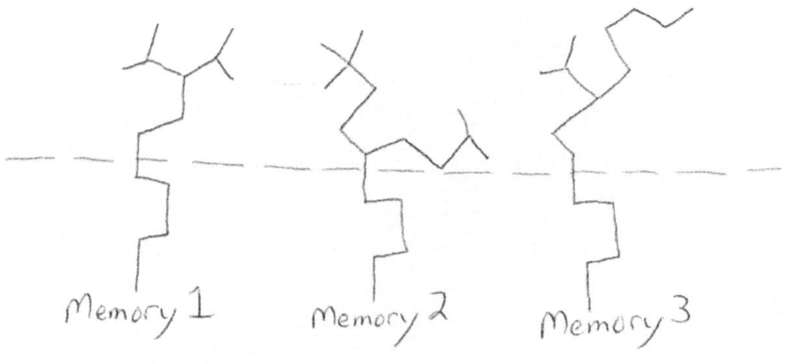


Figure 23: Memory Branches

As you can see in this figure, I have overly simplified neurons into a bunch of corners and lines: The corners represent nerve bodies and lines represent axons. But, as the dotted line dictates, each of the three memories contain the SAME memory trunk. Therefore, if memories are designed like trees, then each neuron can be included in hundreds, if not thousands, of memories.

If memories were like trees, then memory roots are the senses: Seeing, hearing, tasting, smelling, touching, and thinking. Granted, most people don't include 'thinking' in the senses, but I tossed it in there, just to mix things up. The trunk of the tree represents neuronal cord that connects the brain to the senses. And finally, the limbs and leaves represents the massive collection complex neuronal arrangements that are assigned to each memory.

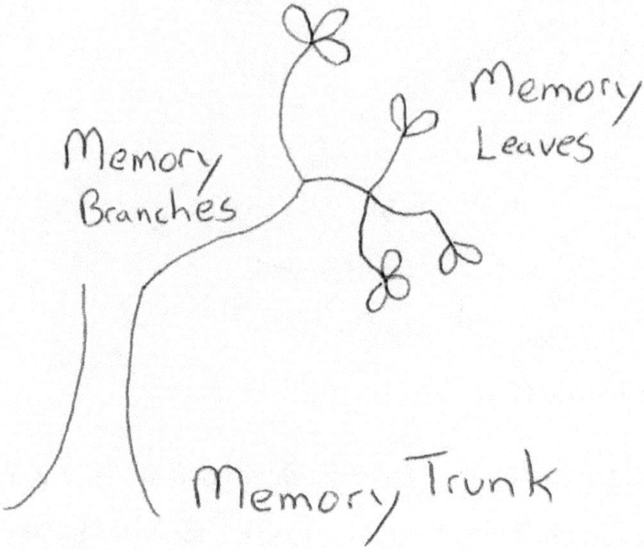


Figure 24: Memory Structure

In this figure, corners represent neurons, lines represent axons, and the leaves represent complex clusters of neurons, which are unique to each memory. Unfortunately, this is where the tree metaphor breaks down because NOT all senses flow through the same region of the brain. Or in simpler terms, each memory can exist as a **collection** of memory trees. Or in more complex terms, consciousness and thought are all intricate weavings of memories and interconnected memories, which are all dependent on how the memories were obtained, the intensity of the event creating the memory (which varies per person), when the memories were obtained, and overall health of the memory branches. Therefore, the old adage "You think like you've thought your whole life" is a very rough approximation of the truth. Actually, I don't think that was an adage. Ooops?

One consequence of memories having the structure of trees is that the FORCE of each neuronal flush dictates the complexity of the memory tree. Or in simpler terms, the **faster** your heart pumps, the **bigger** the memory tree(s). Or in more complex terms, adrenaline also causes the constriction of your myelin sheaths, which increases the pressure of your neuronal flushes; The more pressure to your neuronal flushes, the further those neuronal flushes will travel to create complex memory trees. Or in the simplest terms, a shocking thought can shock your myelin sheaths into a restricted state.

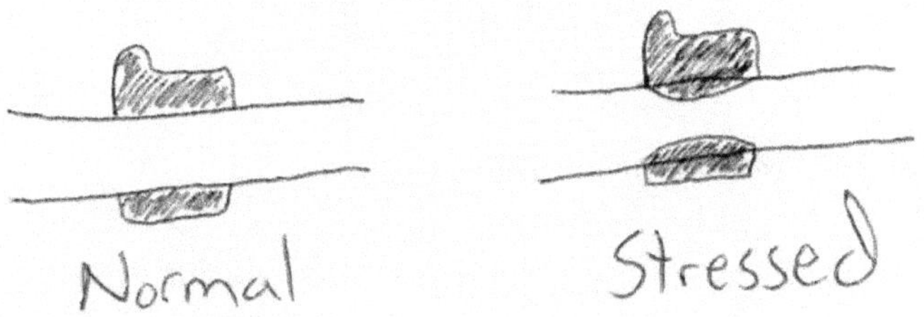


Figure 25: Restrictive State

As a result of myelin sheaths being FOREVER locked into restricted states because of adrenaline, it takes less of a neuronal flush to access memories created during extreme adrenaline states.

Another way to intensifying memory trees is to make them self-stimulating. Or in simpler terms, if a tree branch grows back into the trunk of the memory tree, the branch becomes a memory loop.

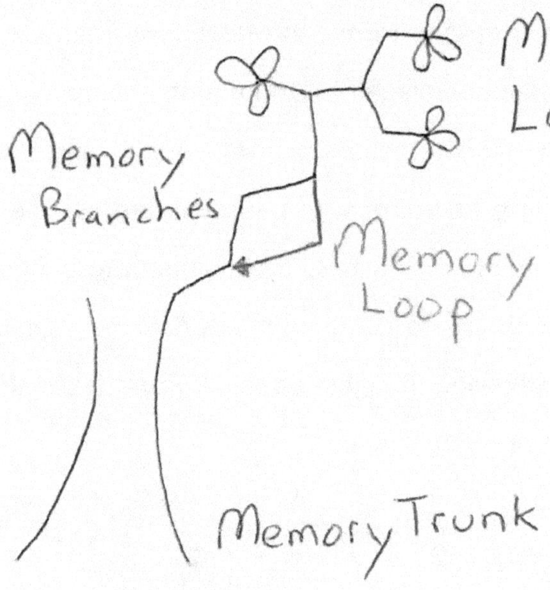


Figure 26: Memory Loop

One unfortunate consequence of forming memory loops is that part of the memory flush is looping back to stimulate another memory flush. Or in simpler terms, you are ALWAYS thinking about that damn thought because of the Memory Loop. Or in the simplest terms, now you know why soldiers are prone to have 'flashbacks'.

With all that shit in mind, nobody knows where consciousness stems from. Maybe consciousness is simply a collection of cyclical loops shooting out memories trees in every direction, based upon stimulus. Whatever consciousness is, unconsciousness is the absence of certain neuronal action. And since memories are designed like trees, it seems probable that dreaming is caused by a jumble of memories trees stimulating other memories trees either through direct stimulation or via proximity. (FYI, memory trees do NOT end in completely ISOLATED neurons. Memory trees are forged based upon the force of a memory flush. Therefore, each memory tree leaf ends in the stimulation of a neuron that is attached to several other neurons.)

In any event, you remember that adage 'Ignorance is bliss'? Well, it is not quite accurate. Any thought can be bliss. I could concentrate on a two-by-four for several decades and assign a joyous meaning to that piece of wood, which would allow me to feel joy every time that two-by-four crossed my mind. Unfortunately, as a result of micro-fluid dynamics, hateful thoughts can also be reinforced, which makes me sad.

In conclusion, I've made an amazingly good argument in support of the postulate that the nervous system runs on micro-**fluid dynamics**. Unfortunately, since science is entrenched by saggy neurons, most of my good ideas seem like shit. But, with a modicum of force, I've been able to flush out another book, which makes me happy enough to poop.

# Cliff Notes

1. 70% of the body is water and the rest is proteins, fat, and/or carbohydrates.
2. Oxidation and reduction of proteins destroys their ability to work, therefore the ULTIMATE function of the body is to CONTROL unwanted **oxidation** or reduction reactions.
3. Mitochondria contain the ONLY structures designed to harness 'electricity' via the Electron Transport Train, which produces ATP.
4. When transmembrane ion channels are stimulated by neurotransmitters, they open up and allow the flow of ions and **WATER** across the membrane into the synaptic bulb.
5. As a result of an increase WATER and ions in the synaptic bulb, water and ions are FLUSHED down the axon, which I like to call the neuronal tube.
6. Movement of water and positive ions down the neuronal tube moves ATP along the neuronal tube, which is the energy that enables proteins to do work.
7. The constriction of myelin sheaths around the neuronal tube, which is driven by the attraction of negative components within the myelin sheath and positive ions traversing the neuronal tube, prevents weak neurological FLUSHES from becoming nerve impulses, the back flow of strong neurological impulses, and the expansion of the synaptic bulbs and/or the neuronal tube.
8. Ions that move down the neuronal tube cause the transmembrane proteins to undergo a conformational change, which opens the transmembrane protein, that causes the influx of more WATER and positive ions to propagate the neuronal impulse/flush.
9. Negative ATP moves down the neuronal axon by following positive ions and provides energy to the transmembrane ion

channels to pump out water and positive ions after each neurological impulse.

10. When the neurological flush reaches the terminal synaptic bulb, pressure and columbic repulsion pushes synaptic vesicles, filled with positive neurotransmitters, to merge with the synaptic bulb membrane and release their neurotransmitters towards the next neuron.
11. The neurotransmitters OPEN the next neuron's transmembrane ion channels, which causes an influx of hydrated ions into the next neuron to propagate the neuronal impulse.
12. The fusing of synaptic vesicles with the synaptic bulb membrane provides reinforcement to the neuronal system by expanding the synaptic bulb towards the next neuron, which insures the next wave of neurotransmitters are released closer to the next neuron.
13. The continual constriction myelin sheaths around the neuron provides reinforcement by decreasing the diameter of the neuronal tube, which increases the PRESSURE of neuronal flushes and the subsequent release of neurotransmitters.
14. Learned behaviors are the result of myelin sheath constriction, an increase in transmembrane ion channels, and the gorging of neuronal Grand Canyons, which facilitates the flow of certain neuronal flushes within a neuron that has several neuronal innervations.
15. Based upon the arrangement of white and gray brain matter, memories mimic tree structure: Trunk, limbs, and leaves.
16. The roots of any memory tree are sensation and/or cognition.
17. PTSD occurs when a loop forms near the memory trunk, which enables the memory to undergo self-reinforcement and/or when extreme Grand Canyons are gorged into nerve cells.
18. Adrenaline not only innervates the smooth muscle surrounding blood vesicles, it also innervates myelin sheaths, which causes an enhancement in neurological flushes, which can result in learning or PTSD.
19. When **negative** electrons surge through the body, they travel through water soluble extracellular matrixes, which sucks out vesicles of **positively** charged neurotransmitters and causes

everybody to ERROUNOUSLY believe that the nervous system runs on electricity.

20. When an ABNORMAL amount of positively charged neurotransmitters are suck out of your neurons by **columbic** attraction, your body freaks-out and you piss your pants.
21. Saggy neurons occur when you get older and result in LESS efficient fluid dynamics, which PUSHES less neurotransmitters towards the next neuron. Examples of saggy neurons are, but not limited to, the following: bent neurons; enlarged synaptic clefts; neuronal regions without transmembrane ion channels; broken myelin sheaths; cluttered nerve cells or synaptic clefts; your mama demanding that she needs grandchildren.
22. As a result of saggy neurons, memory trees can lose limbs, branches, and/or fall down completely.
23. With reference to the arrangement of white/gray matter and the construction of memory trees, consciousness is a roots-**up** process and dreams are a leaves-**down** process.
24. If a memory tree falls in the forest and nobody is there to hear it, does it really gray-matter?
25. Would the 'nervous system' be called the 'system' if people never got nervous?

# *III. The Wobble*

In the beginning, God created the heavens and the Earth…and if you think God had a process, then you're going to hell. Granted, if you don't believe in hell, then you'll just have to live through believing in 'processes' and the hell it provides…if you live in a religious county. Actually, that isn't true. You'll only experience a living hell if you speak out in favor of 'processes' while living in a religious county. Ok, that isn't quite true either. The truth is, your life will be a heck of a lot BETTER if you just keep your mouth shut and believe what everybody else believes. Therefore, if you live in America, you should NOT believe in evolution because science is the devil, unless it is used to wage war in the name of Jesus Christ…or something like that.

If you are wondering, which I'm sure you're not after such a non-confrontational opening paragraph, I totally had to put this book after a book on Quantum Dynamics & neurobiology because ONLY absolute idiots will be reading this non-bestselling theoretical satire mush. Therefore, let me take a moment and pander to my audience: Welcome Idiots! Please do **not** return your seats and/or trays to their original positions because this ain't that kind of book. In fact, this is the kind of book where you take your shirt off and think: 1) Decreasing friction might be the reason why humans still have hair in high friction zones; 2) I really need to stop eating as much and start exercising in order prolong that inevitable myocardial infarction. In any event, please remain seated because reality is about to get a little wobbly: Evolution was facilitated by the degradation of matter into a spectrum of isotopes!

If you feel like vomiting, please, spew away. Honestly, I deliberately forgot to mention that this book might make you spew because I wanted you to blow chunks all over my book, which would make you by my book again...hopefully. (Yeah Capitalism!) In any event, the turbulence you are feeling in every synapse of your nervous system is the result of squishing and squirting, which is based upon the conformation of the proteins. As for the reason for all the squishing and squirting, well, you might be FEARFUL of my first postulate:

> If the conformation of your proteins are dependent the electronic structure of the atoms in your body, then the electronic structure of the proteins in your body can be manipulated depending on the spectrum of isotopes

present in your proteins. And if the structure of your proteins can be modulated by isotopes, then the basis of evolution is the adjustment of genetic efficacy based upon isotopes. Or in simpler terms, if evolution is based upon genetic errors, then the introduction of DIFFERENT isotopes will change protein structures, which will cause more genetic errors and subsequently, evolution.

In any event, that is the premise of this book...in a fucking nutshell. Unfortunately, I gotta slow that knowledge shit down so that people can follow my train of logic. (I think I can, I think I can, I think I can...make you spew again?) Apparently, my brain squishes and squirts much-much faster than the calcium mediated locomotion of my typing fingers. Therefore, Arrrrgh Maties! We're on the same ship but some of us believe in different things. I believe in God and processes. So sit back and wobble in furry if you disagree with me. (FYI, it would be easiest to attack my logical assertions is by pointing out my blatant disregard for separating metaphors with some sort of indentation and/or random spacing.)

# *Chapter 2*

A long time ago in a galaxy far far away, there was an explosion of life. Scientists called it an 'explosion of life'. For some reason, Earth's life-line, which is my term for primordial life, started branching out into new species. Then all those new species started branching out into new'er' species and now you know better than to eat your own poop...I think. On the off chance you haven't figured that last one out, please take it into consideration. Granted, I believe that hereditary characteristics are modulated by external stimuli, mostly because we're not all skinny and beautiful replicas of Adam and Eve, but Darwin didn't know much about physics because Einstein hadn't made physics popular yet. As a result of this, nobody has sat down to think about HOW a slowly changing isotopic spectrum on Earth could have resulted in Evolution. Thankfully, I'm sitting down and I've had more than enough time to think about it.

I don't know if you know this, but ALL isotopes are NOT the same. They have different nuclear components and electronics, which allows them to be separated. Granted, all this knowledge hasn't been used for the improvement of society since we have a global arms race as it pertains to people who refuse to respect one another, but that makes my life much easier. I mean, you gotta look at it relativistically: Nuclear holocaust or pedantic theories of universal human origin...which is worse? Any who, if life developed based upon a SPECIFIC collection of isotopes on Earth, which allowed for the near perfect copying of DNA based upon the structure of proteins, but the 'collection of isotopes' **changed** as a result of a volcano or comet, then how might this shake-up life on Earth? Well, if the precision of DNA replication is based upon the structure of proteins, which is based upon the collective electronic behavior of the 'isotopic spectrum' contained within the environment, then a modification of the 'isotopic spectrum' will change the electronics and structure of the proteins that are responsible for **COPYING** DNA. All of which means, a change in the isotopic SPECTRUM will result in an increased number of genetic mutations that will split species into sub-species, sub-species into monkey-species, and monkey-species into human-species. But all of that is beside the point. The point is: If structure equals function, then structure and function are intrinsically **dependent** on the isotopic spectrum. Or in more complex terms, evolution can occur as a result of protein structural and/or DNA modification. Or in simpler terms, we've all seen the Lego Movie and we know that people are made of little bricks. But, what would happen if the Lego Movie was made with trapezoid bricks instead of rectangular bricks. Or even worse, what if the Lego Movie was made with trapezoid, rectangular, and square bricks? Or worse yet, what if the Lego Movie was made out of bricks that were a half-a-degree off from being rectangular? How would this VERY small angular change be enough to annoy the average person? In any event, the simple answer is: It's complicated.

In conclusion, the structure of the genomic 'house' is based upon the isotopic spectrum present in the building environment. And since the genomic 'house' is built by proteins, the structure of the proteins is dependent on the isotopic spectrum. And finally, since proteins can have hundreds of different 'conformations', evolution is ALSO dependent on protein structure modifications.

# *Chapter 3*

Truth be told, I grew up in a very religious house, which means there was absolutely NO answer to this question: Where did black people come from?

Now I know that that question sounds really racist, but I'm ok with that. I was an ignorant kid and I've said some racist things in my life, but I have the capacity to learn...I think? Unfortunately, I didn't learn the answer to that racist question until after five years of a biology degree and two years of chemistry degree. But before I get to all that, let's talk about racism.

I believe that humanity exists as a spectrum of racists. You may say you're pure of heart, but racism isn't about a philosophical mantra you tell yourself in front of other people. Racism exists out of fear and ignorance. When our ancestors huddled together out of fear, they identified each other based upon looks, which could be the lowest level of cognition. Anybody that looked different was bad because the Earth was pretty harsh. By that I mean, people used to shoot each other over words or bad-looks..."Them there are fighting words partner!" In any event, the need for survival associated with extreme closed-mindedness, gave root to exaggerated hate based upon looks, which has percolated down from generation to generation. But racism can be diminished by education.

When I took organic chemistry in college, I actually read the most of the book. As a result of reading most of the book, I happened upon a small section correlating vitamin D to pigmentation, which postulated that pigmentation was up-regulated as a result of ***too much*** vitamin D synthesis in the skin. (Just in case you've had the chance to take organic chemistry, ***sunlight*** causes a pericyclic reaction in the skin to produce vitamin D, which is fat soluble.) At the time, I thought nothing of this small nugget of knowledge, which was hidden away on half a page near the back of my organic chemistry book. But over the years, I have witnessed some disgusting prejudices based upon phenotypes. People hate other people because of their big noses? People hate people because of their freckles? People hate people because of their hair color? (Actually, I think brunettes are smarter than blonds, but that is beside the point...or is it?)

I thought people were fundamentally good, but in the gray area of self-preservation, I think humanity can get really ugly. Granted, nobody likes to be called a racist, which is understandable based upon Freud's definition of Ego, but maybe the fear-of-discussion will be alleviated if we all agree that everybody is a little racist. I mean, the only way this ignorant-fear will be diminished in society is by education. In any event, if we agree that everyone exists on the spectrum of racism, then it will be easier to talk to people who look different than you, which will alleviate the fear aspect.

So now that I've talked a little bit about the spectrum of racism, here is why most religions should support Darwin's Evolutionary theory...I bet you didn't see that one coming! Quite simply, Darwin's theory of Evolution supports, or at least admits that it is feasible, that humanity started from two people: Adam and Eve. (Aww, does somebody's stomach feel better now?) Truth be told, nobody REALLY knows the exact details of Earth's history. But, the biofeedback system of fat-soluble vitamin D allows for the up-regulation the pigmentation genes based upon amount of vitamin D present during copulation, which creates theoretical feasibly for skin color modulation based upon environmental factors over multiple generations. And before you jump to the conclusion that Adam and Eve were white, let's look at some OBVIOUS points. First, Adam and Eve were naked and in the middle-east, which is land of the boiling Sun. Second, God is really smart. Third, white people seem to congregate in cold places.

Even though you might think the third point might be a little racist, I need to assure you that there is some logic here. For example, WHEN did people traditionally get pregnant in cold areas? The spring. And WHAT season comes before spring? The winter. And WHAT do people wear a lot of in the winter? The clothes. And WHAT does clothes protect from happening in the skin? The synthesis of vitamin D.

As you might be able to conclude, generations after generations of reproducing people that are vitamin D **deficient**, as a result of reproducing after winter, COULD result in the down-regulation of the skin pigmentation to result in paler integumentary system, e.g. skin. Maybe that is why cows were so precious...the vitamin D?

Any who, if someone were to postulate that Adam and Eve were ***brown***, which is STILL the prominent skin tone for humans on this planet Earth, then maybe we all can get along a little bit better. Even if most Religions support Darwin's Evolutionary theory of evolution on a **technicality**, it still is a little bit better than fighting over stupid shit while the universe grows older, our planet gets older, and humans keep hating each other.

In conclusion, I postulate that Adam and Eve were brown. Deal with it as you may. Also, since dietary supplementation has a huge influence on human genetics, which is passed upon to our children via genetic and protein-structure modulation, we all need to be a little more honest about the racist-spectrum before we can move onto something a bit more scientific. Actually, I think I just went backwards with regards to my own logic by writing about science first. Damn it! Now I have to talk about a hypocrisy-spectrum...or not.

# Chapter 4

For those of you who are actually keeping track, we know that obesity is hereditary, vitamin D is fat soluble, protein conformations are DEPENDENT on the spectrum isotopes present on Earth during DNA replication, protein conformations can have a massive effect upon the efficacy of DNA replication, errors in DNA replication can cause evolutionary mutations, and the largest human pigmentation level on Earth is brown. But, how did proteins and DNA become synergistic? Also, how did organisms evolve from using DNA plasmids to supercoiled/convoluted chromosomes? Well, it all has to do with natural oil spills, the oxidative atmosphere, and the electronic migration of electrons away from the Earth's core, which causes lightning.

Human existence is all about compartmentalization. We exist on the Earth compartment. Oil is produced in Earth's crust compartment. Oil can be oxidized to fatty acids when exposed to the oxidative atmosphere. And finally, everybody knows that employees must use fatty acid micelles to cleanse their grubby hands before they return to work. (FYI, soap and fatty acid micelles are synonyms.) Or in more complex terms, electrons must be moving away from the Earth's core to create a reductive compartment that reduces trapped carbon dioxide into heavy hydrocarbons and oxygen, which diffuses into the atmosphere. Then, when oil seeps out of the Earth's crust and is oxidized by the atmospheric oxygen to fatty acids, the first component of life was created. But, how did proteins, carbohydrates, and DNA associate within collections of fatty acid micelles?

Well, since we don't have all the answers...yet...we must DENY ALL the logic collected by science. Wait, how is that going to affect the cognitive growth of the next generation? Damn it, this is never going to work. We have to have all the answers, minus the belief, to believe in the science of evolution such that the next generation is more educated. Umm, wait a minute. Shouldn't I have mocked the intuitive process of God as it correlates to all the goodness that can be created by scientific education? Now my head hurts...I think. Wait, can I use belief to fill in the empty spots of scientific evolution as it pertains to believing in God? On second thought, that shit will never catch on. I guess I'll just talk about the Plasmid Wars.

Before men was men and women were still just a rib in Adam's ribcage, there were magnificent wars being waged by non-cognizant life. Granted, some people define life as being dependent on cognition, but that is beside the point. The point is that the best route to win a war is to confuse your enemy. And the best way for bacteria to confuse their enemies is to give them NON-helpful information. (For some reason, non-synergistic information is just as confusing to bacteria as it is to Christians.) In any event, when a bacteria dies on the front lines of this Plasmid War, the bacteria lyses and releases its plasmid-DNA, which was absorbed by the **opposing** bacteria like a Trojan Horse. As a result of this, the front line of this plasmid war is horrific. Bacteria dying here, bacteria dying there, and bacteria picking up random bits of plasmid-DNA. All of which, allows for a splendid method by which the single celled organisms gained more information. MAYBE, some bacteria developed gyrase protein technology that allowed for the deactivation of foreign DNA plasmids by causing them to supercoil? Wait, if supercoiling foreign DNA plasmids was a defense, then how on Earth did the bacteria begin to incorporate this supercoiled DNA into its DNA program? Hmmm, think Winnie the Pooh, Think! Ah ha, synergy!

Did you know that you are powered by single celled organism? It is true. Inside each of your cells is a mitochondrial organism, which has its own DNA genetic information in the form a plasmid. Is it a coincidence, one of those things God likes confuse the feeble minded with, and/or part of God's process? Well, it could be part of the God process, but that would mean bacteria learned to coexist with different bacteria based upon a synergistic agreement. All of which, would place bacteria on a higher level of cognition than most humans and mean bacteria are living their "lives" in more accordance to God's values? Umm, I really should delete that last question, but it is just too profound to backspace over. Especially if I'm trying to move human cognition forward instead of backwards.

In conclusion, science can be used for good and Evolution is ONLY anti-God if you believe it is anti-God. Also, there is that non-sense about protein wobbles based upon the spectrum of isotopes present on Earth, but that is too much squishy-squish for most people. And finally, I really-really like the philosophy of bacteria, but I hate viruses. Viruses are nothing but lazy tad-bits of genetic information that make my blood boil, especially during the flu season.

# Cliff Notes

1. Isotopes are elements with more or less neutrons, which have slightly different electronic characteristics.
2. Different electronic characteristics can result in different bond angles, strengths, vibration, and/or thermal capacity.
3. Variation in any of the above mentioned bond behaviors results in a 3-dimensional VARIATION when isotopes are incorporated into proteins.
4. Since protein functions are DEPENDENT on protein structure, any 3-dimensional protein VARIATION will modify the protein function.
5. If a protein function is to copy DNA but a 3-dimensional protein VARIATION decreases the efficacy of DNA copying, then errors will occur in each subsequent DNA copy.
6. Errors in DNA copies that makes a species more adaptable to an environment and/or time period, is called EVOLUTION.
7. Therefore, the evolution of the isotopic spectrum intensifies biological evolution.
8. Since protein function is DEPENDENT on protein 3-dimensional structure AND proteins interact with the molecules you eat, DNA expression is DEPENDENT the molecules you eat.
9. Interaction with the external environment causes your body to MAKE different molecules, which modulates the DNA expression in your offspring and increases their chance of survival...hopefully.

# IV. A Viral Spectrum

You might not know this, but genetic mutations have saved your life, millions of times. Unfortunately, the reason why you don't know this is because you were too busy living your 'life' and/or you don't like science. Whatever the case may be, I need you to listen very carefully: There is a spectrum of viral envelopes. Thankfully, there is also a spectrum of ways to prepare eggs, which is the only way for non-scientific people to visualize 'conformational changes' in proteins, which is tied to vaccines and viral envelopes. So let's begin.

If you have ever had the pleasure of aborting a chicken fetus by cooking it, you have observed a protein 'conformational change', which is when the egg proteins turn white. All of which should leave you with two mind-numbing questions: 1) Did humans grown accustom to eating chicken fetuses because of their convenient shell-ish containers? 2) What the hell is a conformational change? Only God knows the answer to the first quandary. As for the second quandary, God knows about that one too, but scientists have been able to learn a bit about conformational changes. I guess the easiest way to describe a conformational change is if you take a sledge hammer to a computer, which will cause a conformational change in the computer and render it absolutely useless. Any who, hopefully you get the mental picture as to how HEAT can destroy a protein by **bashing** the protein to an incomprehensible lump of amino acids. And that ladies and gentlemen, is how scientists first learned how to make vaccines to combat the viral beast.

When I was fifteen, my class-ring got caught on a rafter as I jumped from a garage loft, which tore a large portion of my finger off. Aside from never being in real danger of bleeding to death or losing my finger, but passing out because I was afraid of both, I finally made it to the hospital where the doctor touched the bloody finger to see if I had any nerve damage. Luckily, there was no nerve damage. Unfortunately, my mother was there, the doctor wasn't wearing gloves, and it was during the AIDS scare. All of which, resulted in my mother writing a really-really angry letter to the hospital, which gave me my first memorable science lesson.

Many years later, after a degree in biology and lots of fun classes in graduate school, I finally came to the realization as to WHY heat treated viruses die: Conformational Changes. Or in simpler terms, the proteins in viruses undergo conformational changes because of the heat, which destroys the viral proteins and leaves the virus completely incapacitated. Or in the simplest terms, a virus with several protein conformational changes is equivalent to a computer with several melted wires: Absolutely useless non-functional lump of shit.

With this in mind, things get even more complicated because nobody knows how many conformational changes it takes to effectively kill a virus. And on top of that, nobody knows if different viruses require more or less conformational changes to kill the virus. Quite simply, current virology is a collection of the following things: Isolation of the virus; Controlled growth of the virus; Learning how to COOK a virus just long enough to destroy some internal viral proteins without destroying all of the virus's unique exterior proteins. If the proteins on the OUTSIDE of the virus undergo too many conformational changes, then the human body will NOT be able to make antibodies specific towards the LIVING viruses. So, it is a delicate COOKING process to ensure that all the viral particles are dead and the viral envelope isn't just a lump of **non**-unique amino acids. Luckily, there are some viruses that have broad COOKING ranges that allow for the selective destruction of the interior proteins without destroying all the unique exterior proteins, which the body makes antibodies towards.

Now that we have established what a 'conformational change' is, where a conformational change has to occur to kill a virus and make a vaccine, and that not all viruses are the same, the next logical question is: What the hell is so different about different viruses? I mean, viruses are pretty simple because they are NOT even single celled organisms. Viruses are made up of: Interior proteins, a little DNA, a viral envelope, and exterior viral envelope proteins. That is it. Nothing more and nothing less. But, these little fucking fragments of DNA/proteins can really bitch-slap almost any single and/or multicellular organism. Major bummer if you just realized you're a multicellular organism.

Currently, the most super-terrific method to combating non-vaccinable viral beasts is to feed the viruses modified DNA/RNA molecules, which inhibits the virus from copying its DNA/RNA. Or in much simpler terms, <u>**modified**</u> DNA/RNA molecules cause a massive paper jam when the virus is trying to <u>***copy***</u> its viral information. Modified DNA/RNA molecules like acyclovir, which is used to treat herpes, is an example of a drug that causes a paper jam in the copying of the virus's DNA/RNA information. Unfortunately, inundating a multi-cellular organism with **funky** DNA/RNA often causes lots of problems. It is akin to the treatment of cancer, which is very problematic. Therefore, finding a vaccine is the easiest way to treating a systemic viral inundation. Unfortunately, not all viruses are the same.

In as much as some people would like to think that virology has reached its peak, I totally beg to differ. We have reached a precipice of human knowledge and it could change our chances of survival in this harsh universe. Here is what we know:

1. We know which modified DNA/RNA molecules inhibit viral copying.
2. We know that some viruses are different than other viruses.
3. We know that the ***HEAT*** scrambling viral proteins is ***ONE*** method to creating vaccines.
4. We know the human body can make trillions upon trillions of antibody-variations via controlled genetic mutation.
5. We know that vaccines have forever changed human history, evolution, and the level of cognition.

All we have to do now is take the next logical step in combating the viral beast.

If we imagine that there is spectrum of viral envelopes stabilities, then we can invent gentler methods to killing viruses. Granted, I'm sure the pharmaceutical industry has tried everything from gamma radiation to ultraviolet light to kill viruses without destroying the unique exterior proteins of a virus, but maybe it is time for a much-much gentler approach?

# Chapter 2

In order to debase the current modular theory behind virology, let's use our imaginations. And if you don't mind, I think I'll go first. I imagine viruses are like little books: DNA/RNA are the pages in the book; Internal proteins are the glue that binds the book together; Book cover is the viral envelope; Design on the book cover is the unique exterior proteins of the virus. Unfortunately, some viruses have ***hard*** book covers and some viruses have ***soft*** book covers. I'll give you a second to calibrate your imagination to my settings...

Next, let's imagine a Nazi book bond-fire. For those of you who haven't had the privilege of witnessing such an event by TV or in real person, here is a nugget of knowledge: Hard book covers are slightly more fire resilient compared to soft book covers. Or in simpler terms, hard-cover books burn SLOWER and soft cover books go up like a Roman candle. (Side note: The Chinese invented fireworks. Also, the Romans perfected public execution as a means to keep the masses from rebelling. And finally, I don't know where the term 'Roman Candle' comes from.)

With all that in mind, minus the side note, let's remember that the human body is very-very superficial. By that I mean, the human body DOES judge a book by its cover. If a book has an unusual book-cover, the human body will make anti-body towards that uniquely-odd book-cover. If a book has a cover that is like all the other cells in the human library, then body will NOT make an antibody towards that book-cover...usually.

Hopefully your imagination is severely confused. Should you be thinking of books, Nazis burning books, public execution, Chinese people making fireworks for America's 4th of July, or the human library, which contains weird sections like: S-conformational amino acids are for losers? Actually, when you are confused as hell, this is when the mind has the amazing propensity to concoct something new. In any event, you should be imagining how to COOK the Polio virus just long enough to burn some of the pages from the book without destroying the unique book-cover, which allows the human body to make an antibody towards the Polio viral book-cover. Also, you should be imagining how some viruses have tissue-paper book-covers that can rip when they are exposed light and/or air.

We know that some viruses can survive in the dried human blood for weeks; like the hepatitis viruses. We know that some viruses can survive for a couple days in human sneezes; like the common cold. We know that some viruses die almost immediately when exposed to an oxidative air environment; like the HIV virus. Or in simpler terms, hepatitis virus has a hard book-cover, the common cold virus has a paper book-cover, and the AIDS virus has a tissue-paper book-cover. All of which means, it is hard as hell to burn a couple pages of the AIDS viral book without completely destroying the tissue-paper book-cover. Or in extremely difficult and annoyingly complex terms: There is a spectrum of viral envelopes!

We know that the conformation of the proteins on the exterior of the viral envelope are very sensitive to the structural integrity of the viral envelope. Therefore, when a viral envelope breaks, it destroys the conformational complexity and uniqueness of the proteins on the viral envelope, which the human body targets with antibodies. Or in simpler terms, when a virus's envelope cracks, the structurally unique **exterior** proteins on the viral envelope become a jumble of non-specific amino acids. Or in the simplest terms, proteins on a cracked viral envelope do NOT look like the proteins on a NON-cracked viral envelope.

And so ladies and gentlemen scientists, this has been the conundrum for several decades: If a viral book-cover burns\breaks before being able to burn a couple pages in the viral book, then a vaccine ***cannot*** be made against the delicate virus. All of which, results in doctors inundating patients with funky DNA/RNA molecules to prevent the virus from replicating. BUTT, what if we stick some gum in the viral book? Ahhhh ha! Curing AIDS with a gum analogy...totally inappropriate unless you know my second option was referencing a sticky-page scene from one of Seth Rogan's movie.

Now that our imaginations are on the same sticky page, what does all this mean? Well, scientists have known all about how toxins can cause DNA dimerization for decades, but people are still happily smoke cigarettes every day because the human body has the propensity to repair DNA damage. But we ain't talking about the human body! We are talking about a fucking short viral book that has absolutely NO way of correcting damaged/dimerized DNA. Ahhhh ha! You see where I'm going...don't you?

Instead of burning a viral book to destroy a couple pages, why don't scientist make modified DNA/RNA molecules that will polymerize and glue several pages of the viral book together? All of which, should enable scientists to create DEAD viruses without even TOUCHING the viral book cover! Yeah science! Wait, does anybody know about polymerization, which types of polymerization can be started gently, and where we can place these functional-polymerization equivalents on DNA/RNA such that it doesn't interfere with the formation the virus? Damn you God for making this so complicated!!! (Just kidding God, you know I love you? I was only trying to create dramatic tension within this scientific narrative;-)

Hopefully you haven't stopped reading as a result of my blasphemous outburst because we are just getting to the area of my educational expertize...Organic Chemistry! I love OChem...Oh, Oh, OhooHo, the Chemistry stuff. Wait, where did all the readers go?! Please come back! I promise I won't start talking about functional groups, anisotropy, and/or the Backstreet Boys. (Boy, this is going to be difficult to talk about if I can't talk about organic chemistry.)

For those of you have who haven't had the pleasure of laying your eyes on an artist's depiction of an antibody, then you truly haven't lived. I mean, you might be living, but you are truly living if you haven't appreciated what the body does to keep you alive every day? First, there is the long protein strands of the antibody, which is like a pair of never ending legs. Next, there are the short, but very strong, protein strands, which are like a pair of hands stroking those long sexy protein strand legs. Finally, there are all these hydrophobic, hydrophilic, and chemical bonds that make these protein strand legs wrap around each other in an orgy of structure and function. And the beauty doesn't stop there. You see, as a result of the world NEVER staying the same, which some people call evolution, viral book-covers can change. As a result of the viral book-covers changing, the human body has to build different antibodies. As a result of needing different antibodies to combat different viruses, our DNA is manipulated to produce different antibodies. As a result of our DNA having to change, without developing cancer, antibody-cells have to masturbate, which is a great metaphor to help the next generation admire the beauty of their bodies…no pun intended.

You might not know this, but your children are NOT absolute copies of you or your significant other. In fact, the combination of sperm and egg DNA is caused by Holliday Junctions, which usually occurs around the Holidays. All of which, creates a progeny that is a unique mixture of the sperm and egg DNA. Unfortunately, all this massively complex DNA information cannot code for every possible viral book-cover that will ever exist. I mean, DNA codes for a lot of information, but humans would be a whole lot fatter if our DNA also contained the information for EVERY possible viral book-cover. That is why antibody-cells are "jacking it in San Diego" as the kids on South Park are saying these days. Or in different terms, antibody-cells are DNA casinos. When a new viral book-cover appears, antibody-cells shuffle some antibody-DNA-components and then they deal out a new antibody. Unfortunately, the antibody-cell-DNA-casino is relatively slow, which is why people get sick and then better. Granted, you can help your antibody-cell-DNA-casino by eating healthy and staying hydrated, but who's got time for that, right?

In conclusion, the body is really AMAZING at keeping us alive by making novel antibodies. Also, there is a spectrum of viral envelope stabilities. And finally, all we have to do to is invent a way to 'glue' a couple pages of the little viral books together such that our body can make a vaccine towards those 'dead' viral books. Pretty simple, right?

# Chapter 3

So where should I begin when I've had so many ideas about how to combat evil viral beasts? Well, I guess I should start with the worst idea and make my way to the best idea. In the event you want my ideas in reverse order, then read this chapter backwards as it pertains to the paragraphs, not the actual sentences.

<u>Idea number one: Bad-bad-idea</u>

When I first started thinking about decreasing the viral load as it pertains to an infected individual, I envisioned a two pronged approach. First, there would be the traditional antivirals that lower the viral load in the blood stream. Second, there would be a super complex filtration machine that would bind excess viruses and decrease the viral load even more. Granted, this highly technical filtration machine would need a crap load of research as it pertains to recirculating the blood while actively allowing the diffusion of the very-very small viruses via an osmosis type setup, which would allow the viruses to be exclusively exposed to polymer bound horse-antibodies, but then there is the problem of horse-antibodies. As a result, this idea was total horse crap.

Idea number two: Sort-of-bad-bad idea

My next idea was exactly the same as the above idea expect instead of horse-antibodies, I'd use pig anti-bodies. Just kidding. Instead of horse-antibodies, I thought about irradiating the filtrate with some sort of ultraviolet radiation to destroy all the viral particles. Unfortunately, this idea still includes several years of research as it pertains to a very complex osmosis blood machine. Not to mention, I don't think most AIDS patients could afford to be attached to a machine until they are cured.

Idea number three: A Bad idea

My next idea was to continue my former work into the synthesis of naturally occurring antiviral molecules within green tea. Except for the fact that this was one of Thomas Kinstle's ideas and I no longer have access to laboratory, it was actually a really good idea. You would not believe how terrific green tea can be for you. I bet someone will find a useful analog of one of those green tea molecules…someday.

Idea number four: An Idea

My next idea was the photo-polymerization of modified DNA/RNA molecules within viruses, which was kind of inspired by Vladimir Popik's work on enediynes. Unfortunately, as a result of realizing that there are a spectrum of viral envelopes, I'm not quite sure if there is also a spectrum of conditions that viral envelopes are sensitive to. For example, some viruses might be sensitive to oxidation, others might be sensitive to light, and others might be sensitive to different pH levels. Also, it would take a lot of work to determine which precursor functional groups would NOT bother viral production. And on top of all that, different viruses would have different DNA/RNA conformations that may or may not allow for the proper placement of polymerizing functional groups.

Idea number five: A good idea

As a result of the last idea's snafu as it pertains to the DNA/RNA conformations that allow for polymerization, I came up with the idea of Click Chemistry. Actually, I didn't come up with the idea of Click Chemistry, but I learned about it as a result of working under Stephen Bergmeier. Quite simply, alkynes react almost perfectly with azides in a 1,3-dipolar-cycloaddition to make substituted triazoles. It is an amazing reaction and it might be perfect for the mild polymerization of viral DNA/RNA. All you would have to do is make DNA/RNA molecules with an alkyne OR azide, allow the virus to reproduced and lyse the host cell, then treat the viruses with corresponding di-alkyne OR di-azide aliphatic cross-linker to cause polymerization of the viral DNA/RNA. Granted, it wouldn't technically be polymerization, but who cares. It would only take two or three cross-strand linkings of aliphatic di-triazoles to completely incapacitate/kill the viral particle. And since both the alkyne, azide, and aliphatic linker are all neutral species, they can EASILY diffuse across the viral envelope. Granted, somebody will have to figure out level of treatment of the virus with the di-alkyne or di-azide aliphatic cross-linker and if the aliphatic cross linker would need a hydrophilic functional group, but this seems like the best idea yet. Also, I suggest that an actual polymer bound azide or alkyne be used to scavenge the excess di-alkyne or diazide aliphatic cross-linker, which can be easily filtered off. In any event, special considerations might have to be made if it is discovered that the HIV virus is sensitive to oxygen, pH level, or light.

Idea number six: The best idea

Whatever fucking works.

In conclusion, I postulate that there are a spectrum of viral envelopes: Some very strong and some very weak. Also, I postulate that Nazis and book burnings are bad. And finally, I postulate that Click Chemistry is the best route to gently gluing a few pages of the tiny viral book together such that the human body can make antibodies towards the "dead" viral book-cover.

# Cliff Notes

1. Protein function is DEPENDENT on protein structure.
2. A change in protein structure is called a CONFORMATIONAL CHANGE.
3. Denatured is the term used to describe a conformational change that renders a protein NON-functional.
4. Viruses consist of external proteins, internal proteins, lipid envelope, and some DNA.
5. When the body makes an antibody towards a virus, it targets UNIQUE **external** protein conformations.
6. Destruction of a virus's lipid envelope results in the denaturing of external proteins.
7. There is a spectrum of viral lipid envelopes, which makes some viruses more resilient to the external environment.
8. Vaccines are traditionally fabricated by denaturing internal viral proteins with heat.
9. A vaccine cannot be fabricated if the lipid envelope is destroyed before the denaturing of internal viral proteins.
10. Viruses do NOT have systems to repair damaged DNA.
11. Antivirals are modified DNA molecules that disrupt viral DNA copying.
12. Click-Chemistry is genre of chemistry that is built upon reactions that are 99% efficient.
13. It is theoretically possible to create modified DNA molecules that do NOT inhibit viral replication, but undergo Click-Chemistry to incapacitate viral DNA.
14. Incapacitation of viral DNA without damaging external viral proteins should provide a very mild method to producing dead viruses, which can be used to make vaccines.

# V. Magnetically Enhanced Synthesis

Things inside NEGATIVE magnetic fields are weird. Actually, things are only weird when you think about it, apply Quanta Dynamics, and imagine everything in terms of COLUMBIC interactions. Granted, most people don't do that, but that is why I decided to venture into this cluster-fuck...albeit briefly.

The reason why I call it a cluster-fuck is: It is a tapestry of intertwined theories, which I've postulated over the course of a couple books. (Not this collection of books per se, but you know, the other books I've flogged society with?) In any event, here is a quick review of those afore mentioned postulates. (Honestly, I've written so much that I'm not quite sure which postulates I've published. On an unrelated note, I postulate that if I'm MEAN to society, which is a bitch sometimes, then society will like me more...or something like that. In any event, prepare to be flogged?)

In accordance to my postulates that negative magnetic energetic quanta are degradations of negative electrons, which are stimulated to decay by running into negative thermal energetic quanta, super conducting magnets are complicated. The circular movement of electrons about the super conducting matrix results in the directional distortion of perpendicular atomic orbitals and their energy containment regions therein, which contain negative thermal energetic quanta. As a result of this, the electrons attached the elements within the circular magnetic matrix, which are arranged perpendicular to the movement of the external electrons flowing around the super conducting magnet, run into negative thermal energetic quanta, degrade, and release negative magnetic energetic quanta perpendicular to the movement of the electrons flowing around the super conducting matrix.

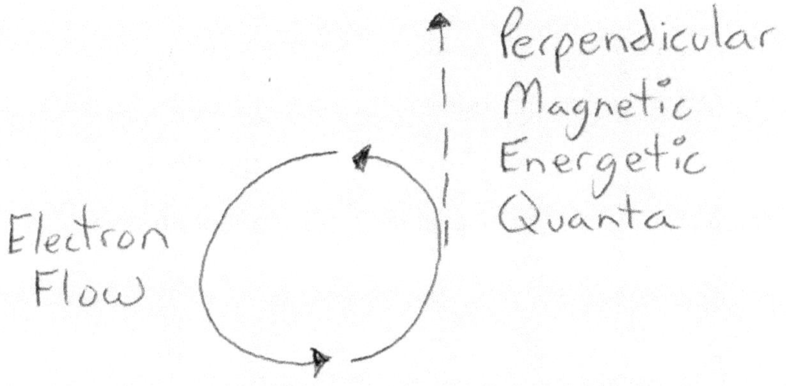


Figure 1: Perpendicular Decay

As you can see in this figure, I did NOT attempt to draw perpendicular atomic orbitals, which are being distorted by millions of external electrons flowing about the super conducting magnet. Or in the simplest terms, I suck at drawing.

With those clarifications in mind, let's turn to the behavior of the molecules that scientists place within these NMR machines. The molecules in the NMR are being saturated with negative magnetic energetic quanta. As a result of this massive directional influx of negative energy towards these atomic orbitals and protons, which are also releasing negative magnetic energetic quanta, two different energetic states are created: The higher energy state is when a proton is releasing its negative magnetic energetic quanta TOWARDS the non-valence atomic orbitals; The lower energy state is when a proton is releasing its negative magnetic energetic quanta AWAY from the non-valence atomic orbitals. (FYI, this explanation is not absolute and is dependent on atomic thermodynamics, which I'll explain in subsequent books.)

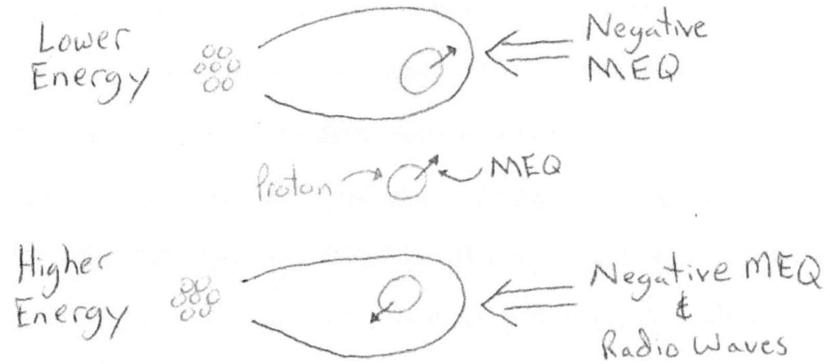


Figure 2: Proton Partition

When a proton within a strong magnetic field is irradiated with specific radio wave frequency, it vibrates the proton from a lower energy arrangement to a higher energy arrangement, which slowly decays by procession and releases a specific radio wave that can be detected and plotted by the NMR machine.

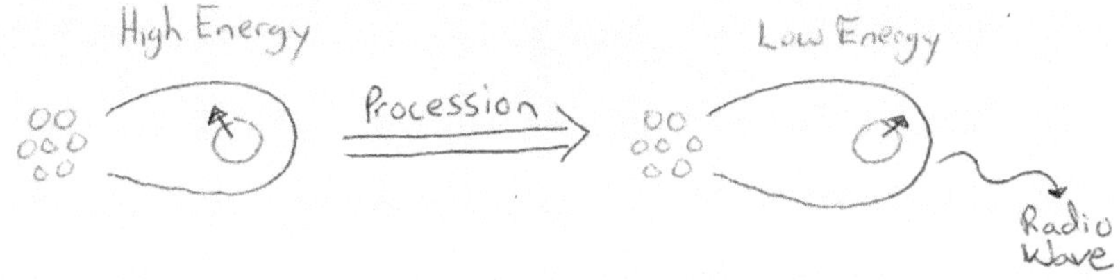


Figure 3: Proton Procession

After carefully ponding the above figure, which I'm sure everybody does without implicit instruction, it is important to note that the energy fluctuating through the NMR machine is similar to the energy that creates Earth's magnetosphere, but the energy within the NMR machine is denser. Therefore, compounds in the NMR machine behave differently. Not as differently as compounds and matter act when thrust into a black hole, which destroys all chemical bonds and most matter, but the energy within the NMR machine DOES make the molecules behave differently, which means the energy within the NMR machine is MODULATING the chemical and electronic behavior of the atomic orbitals. Unfortunately, before I can extrapolate these postulates into Magnetically Enhanced Synthesis, I NEED to give an annoyingly vague, yet specific, review of organic chemistry.

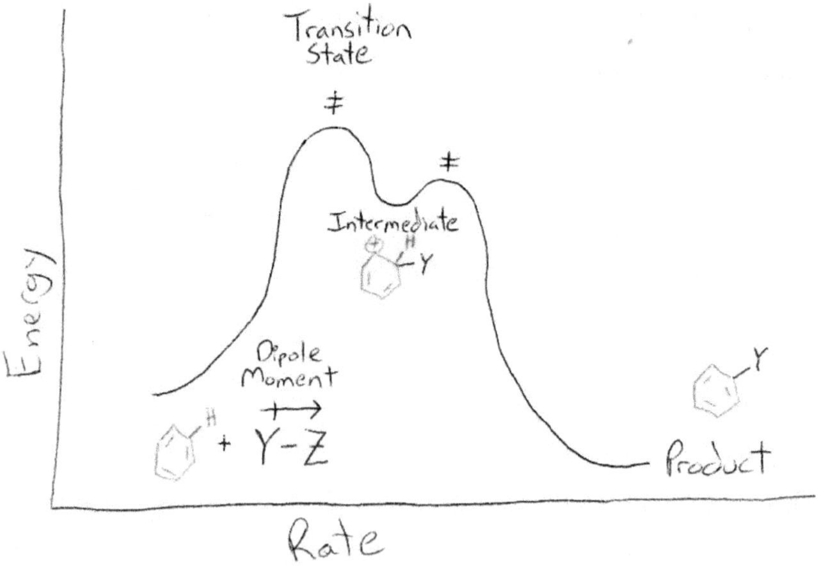


Figure 4: Energy Diagram

As you can see in this energy diagram of a generic electrophilic aromatic substitution, there is a dipole moment, two transitions states, an intermediate, and a few other lines, letters, and words. All of which, is pretty straight forward to any organic chemist.

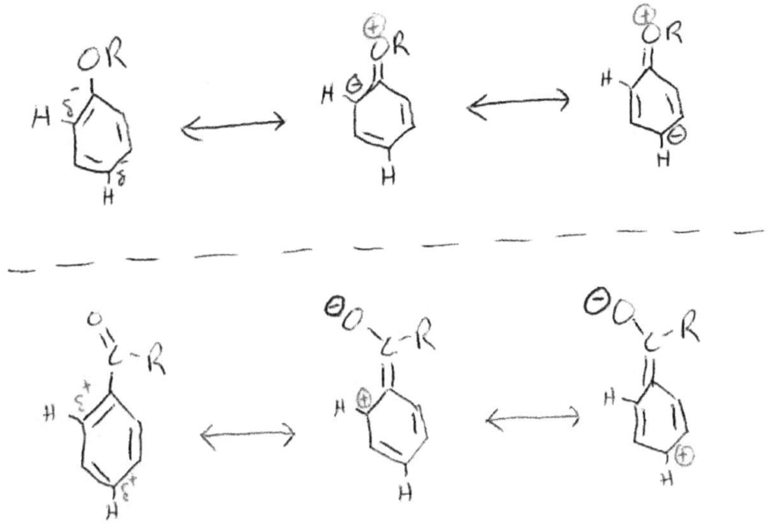

Figure 5: Resonance Structures

As you can see in this diagram, more lines, letters, and words are used to review the concept of resonance structures. I guess, I could be more specific, but I've personally seen students that are crippled by these concepts, even after taking organic chemistry for a year. In the event that those two figure resulted in some mental recognition, let's move on.

So the premise of Magnetically Enhanced Synthesis is quite simple: Inundating compounds with vast amounts of negative magnetic energetic quanta modulates the electronic and chemical uniqueness of certain atomic orbitals, which is basis of all synthetic utility. Or in different terms, Magnetically Enhanced Synthesis is the ability to TUNE dipole moments to adjust chemical bond reactivity.

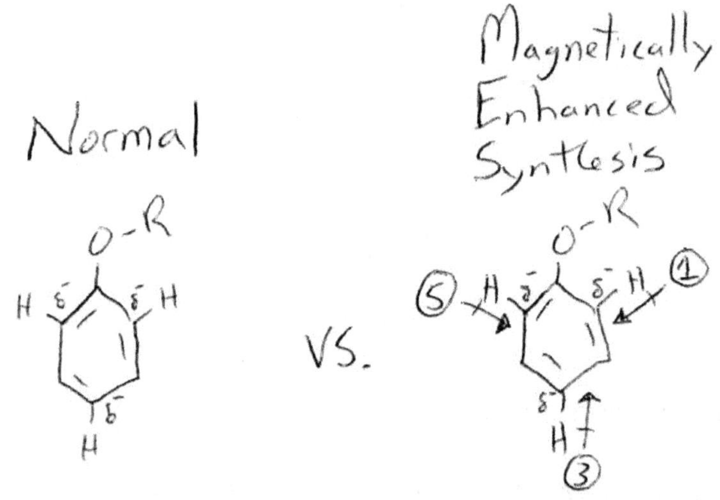


Figure 6: Magnetically Enhanced Synthesis

As you can see in this figure, hopefully, the 'Normal' compound contains three partial negatives, which is the result of resonance. But, in the 'Magnetically Enhanced Synthesis' compound, a SCIENTIST can selectively ADD a dipole moment to position 1, 3, or 5. (FYI, the molecule in the figure is symmetric and position 1 and

5 are identical, but if there was a meta-substituent, things could get really fun.) Since ALL chemical reactions are based upon electronics, a SCIENTIST could fine-tune a structure to make certain spots more reactive with dipole moments. But the fun doesn't stop there! By that I mean, I have describe the METHOD of this dipole moment and its ability to modulate chemical reactions...albeit in the next chapter.

In conclusion, Magnetically Enhanced Synthesis will enable to SCIENTISTS to electronically differentiate **components** of a molecule WITHOUT performing expensive chemical modifications. Unfortunately, Magnetically Enhanced Synthesis will NOT work in every reaction that involves protons, but it still has the capability to save Pharmaceutical Companies billions of dollars by increasing the yield in a few reactions. (Honestly, I totally don't want to go into the goodness or badness of Pharmaceutical Companies. Their endeavors have saved my life more than a couple times. Therefore, let's just leave it at that.)

# Chapter 2

To understand Magnetically Enhanced Synthesis, all you gotta realize is: Protons have ASSes. Actually, it is a little more complicated, butt I don't want to scare you with my mental flatulence. Therefore, let me try and make this as simple as possible.

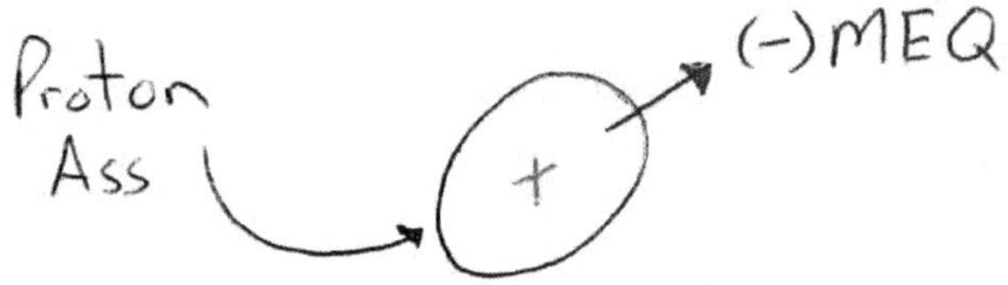


Figure 7: Proton Ass

Whichever side of the proton is not squirting out a negative magnetic energetic quanta [(-)MEQ], is the proton ass. As for the reason why protons have asses, well that is quite simple: Protons gets sore from sitting on negative atomic orbitals.

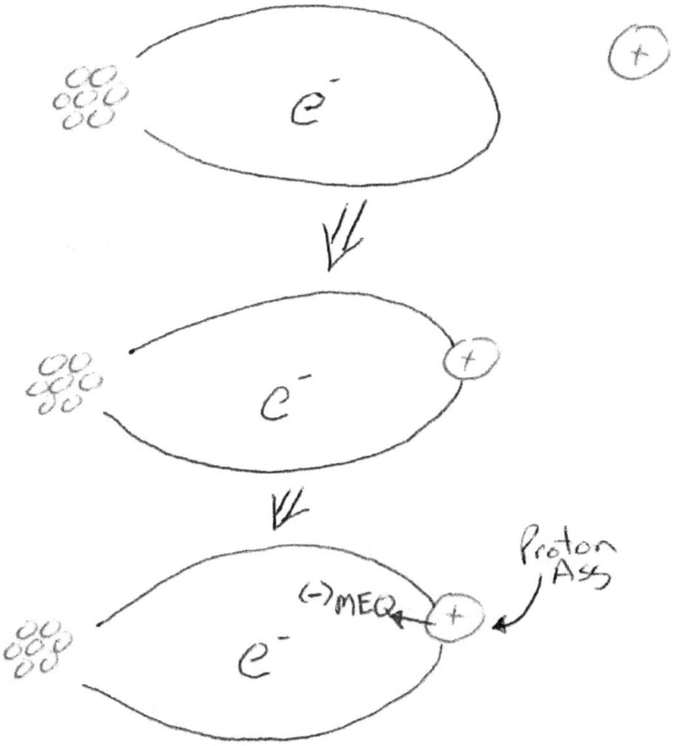


Figure 8: Creation of a Proton Ass.

As you can see, the proton exists without an ass until sits down on a negative atomic orbital, which contains lots of negative plasma. As a result of all this negative plasma degrading part of the positive proton, the proton releases some negative magnetic energetic quanta such that it can maintain its positivity. And the moment after the proton fires-off its first negative magnetic energetic quanta, the proton develops a massively big positive ass. (FYI, protons don't NEED a lot of momentum, like electrons, to be stimulated to decay and release negative magnetic energetic quanta because: 1) Positive protons attract negative plasma whereas negative electrons repulse negative plasma; 2) Positive protons are MASSIVELY expanded because they exist in this negative branch of the universe whereas electrons need to be transported to the magical land of pixie dust and positivity, e.g. positive atomic nuclei, to be expanded.)

One problem of protons firing-off negative magnetic energetic quanta and developing positive asses is: The conservation of energy. You see, protons are attracted to all the negativity within atomic orbitals, but the protons understand that a co-dependent relationship will result in mutual destruction. Therefore, the proton tries to protect its positivity by releasing some negative magnetic energetic quanta, which does a great job of PUSHING the negative plasma away from the proton. Or in simpler terms, sexy protons have to carry large cans of negative-MEQ-pepper-spray otherwise the sexy protons would be destroyed by all the negative plasma contained within an atomic orbital.

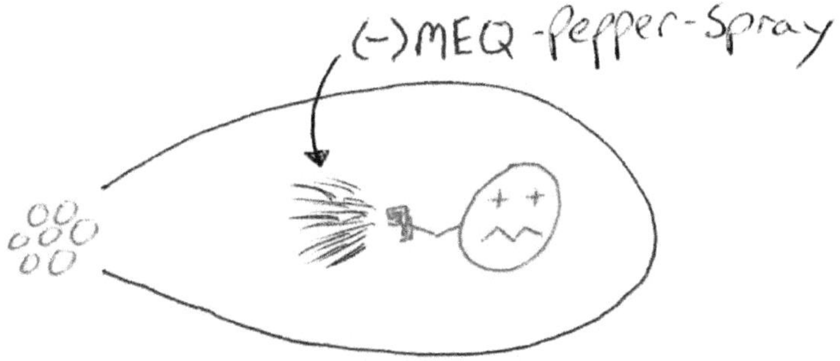


Figure 9: Bitches gotta protect their asses

Unfortunately, protons are like Sandra Bullock in that Gravity movie: Lost in Space. (FYI, I don' think Mrs. Bullock is a Bitch.) Therefore, as soon as the sexy proton starts shooting its negative-MEQ-pepper-spray, it starts to spin around, which causes its positive ass to meet up with all the negative plasma in the atomic orbital.

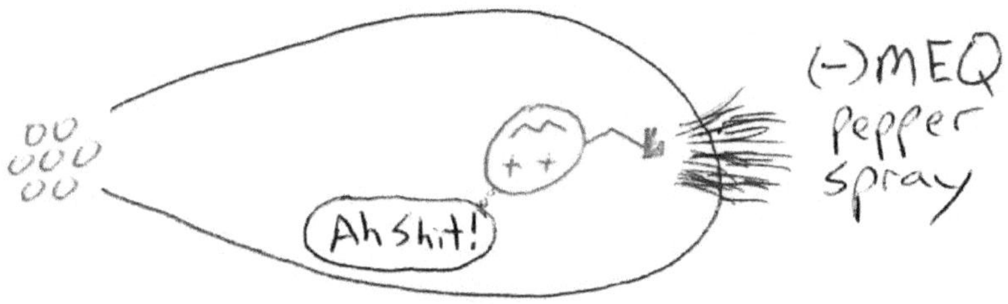


Figure 10: Negative-MEQ-pepper-spray in Space

Butt as soon as the proton spins around and exposes its positive ass to all the negative plasma within the atomic orbital, the negative plasma begins to degrade the proton's positive ass, which results in the magical ass-arm switch-o-rue.

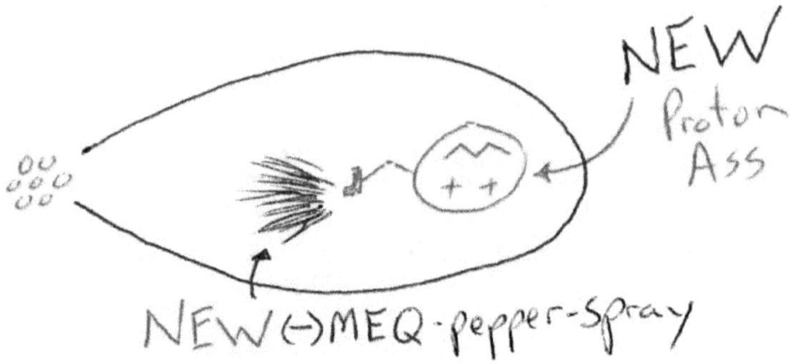


Figure 11: Magic Booty Movement

And so ladies and gentlemen, this is the life of a proton. (Granted, it's not magic that causes the proton arm and ass to switch places, but that is beside the point.) The point is that a proton has to be eternally doing the arm-ass switch-o-rue to protect itself from negative plasma. Unfortunately, SCIENTISTS have devised a way to make protons ALWAYS show their positive asses.

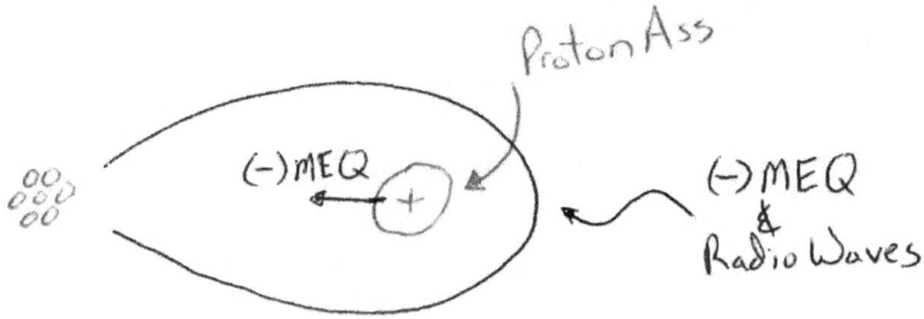


Figure 12: Scientists Gone Wild

Hopefully, most people are not turned on by this grotesque display of positive proton ass-ery, but I know some very basic atomic orbitals that spend their whole lives cruising for a piece of positive proton ass like this. Or in more complex terms, check out this next figure.

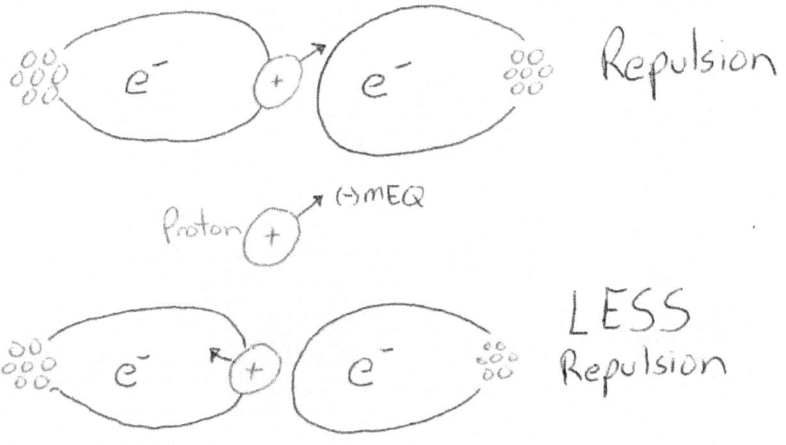


Figure 13: LESS Repulsion

So where does theorizing leave us? Well, if a scientist can make a proton have LESS repulsion towards an incoming atomic orbital and MORE repulsion between the proton and its current atomic orbital HOME, then a scientist can MAGNETICALLY change the acidity of proton...WHICH IS MASSIVE!!! No seriously, it is fucking massive. By simply running a reaction in an NMR while constantly irradiating a SPECIFIC proton with radio waves, a scientist can MODIFY the acidity of the proton, which can increase selectivity and yields while saving money!

In conclusion, protons have positive asses. Also, scientists can control the direction these proton asses within NMR machines. And finally, controlling the direction of a proton ass can affect the acidity of the proton...among other things. (Wink, wink, nudge, nudge, next chapter.)

# Chapter 3

In as much as I put the cart in front of the horse, by describing the dipole moment created as a result of the proton's release of a negative magnetic energetic quanta, I feel the as if I should take a step back and explain 'normalcy', which should have proceeded both the cart and the horse. Granted, normalcy is the behavior of protons within the Earth's magnetosphere, but that is beside the point. The point is: Normalcy is an average.

For those of you who don't know this, protons are fucking crazy. They hang out with these negative atomic orbitals, which are trying to kill the positive protons. As a result of this, the protons are continually pepper spraying the atomic orbitals with negative magnetic energetic quanta. I guess you could call it a love-hate relationship, but it is a little more complicated than that. Unbeknownst to the protons, the atomic orbitals are in a committed relationship with all the other atomic orbitals attached to the same atom AND all the other atomic orbitals within the same molecule. (BTW, atomic thermodynamics will be explained in later books.) All of which, gets extremely complicated and unique. As a result of ALL of this uniqueness, some protons are lured further into these abusive atomic orbitals, which changes the way the protons spin around as they release their perpetual negative pepper spray. Or in simpler terms, the average normalcy of proton rotation, location, and release of negative magnetic energetic quanta pepper spray is not only grossly complicated, but it is grossly complicated by the external negative magnetic field.

With this pseudo explanation of proton behavior in mind, the take home message is: There are a lot of factors to HOW MUCH positive proton ass is displayed to the average viewer, i.e. every other electronic entity not attached to the same molecule. And just in case you didn't know, HOW MUCH positive proton ass is shown to the average viewer, determines a lot of things. For example, one of those 'things' is how fast molecules travel through a matrix of silica gel. Or in simpler terms, the more positive proton ass that is shown to the silica gel, the SLOWER the proton, and subsequently the molecule that the positive proton is attached to, will move though the silica gel. Or in the simplest terms, silica gel loves positive proton asses. Or in more scientific terms, the more acidic a positive proton, the slower it will move through silica gel.

With all this in mind, it should be simple to see that NMR chromatography will be the wave of the future because this will allow researchers to modulate SPECIFIC protons on SPECIFIC molecules to adjust the RATE by which they proceed through a silica gel column. Or in much simpler terms, pharmaceutical companies spend TRILLIONS upon TRILLIONS of dollars purifying EACH intermediate of EACH drug they make...so Cha-Ching! Or in the simplest terms, I know what some corporation-peoples what for Christmas!

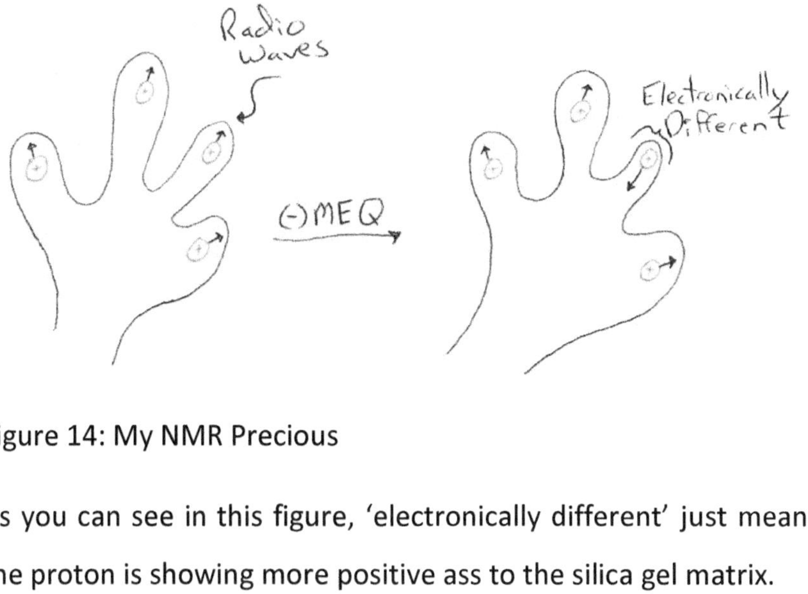


Figure 14: My NMR Precious

As you can see in this figure, 'electronically different' just means the proton is showing more positive ass to the silica gel matrix.

In conclusion, all you gotta do is drop a **normal** silica gel column, HPLC, or GC column into a NMR and constantly irradiate one **OR** more protons such that the electronics of the molecule and the Proton-Ass-Factor are modulated. Upon molecular modulation, which will result in a DIFFERENT interaction of the molecule with the solid-state column matrix, the molecule will display a varied retention rate and will be easier to isolate...Cha-Ching! And finally, a Chiral NMR machine would probably make a corporation-person blow their load on Christmas morning.

(Prices and availability of a Chiral NMR machine is based upon the price of my next book, ***Imperfect Perfection***. Until such time, NMR enhanced synthesis and purification will only work on diastereomers and other non-enantiomer compounds. The author is also not responsible for ANY illegal drugs procured by NMR enhanced synthesis or purification, but reserves the right to enjoy such high-class drugs in the comfort and confidentially of his own home...Amen.)

# Chapter 4

**Readers Beware! EXTREME organic chemistry will follow!**

There is this reaction called Electrophilic Aromatic Substitution (EAS) and it was pretty awesome!  First, there is the resonance structure of an aromatic ring based upon the functional groups attached to the ring.

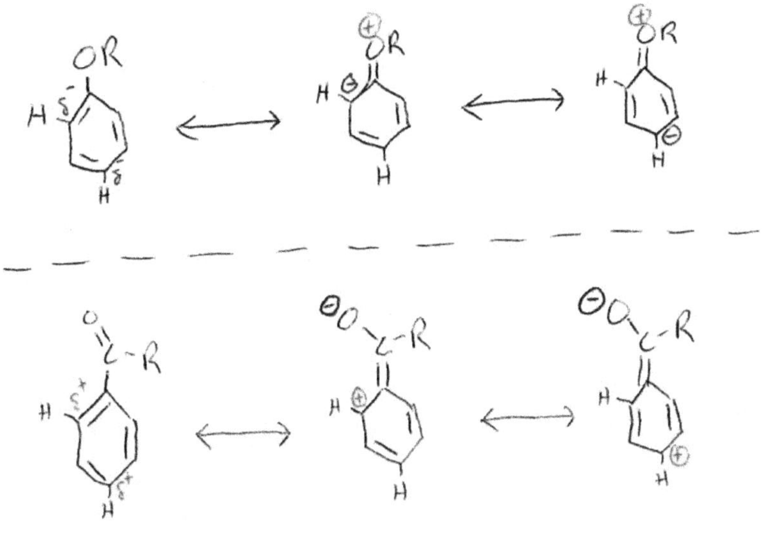

Figure 5: Resonance Structures

Hopefully, you are blissfully unaware that electron donating functional groups, notated in part 1 of Figure 5, direct Electrophilic Aromatic Substitution (EAS) to the Ortho or Para positions. And, you should feel absolutely ashamed that you don't know that electron withdrawing functional groups, notated in part 2 of Figure 5, direct EAS to the Meta position. All of which, is the result of...wait for it...wait for it...Resonance Structures. As a result of Resonance Structures, different carbons have different amounts of negativity. And since the Rate Determining Step of the reaction is dictated by the amount of negativity residing on a carbon, the more negativity on a carbon will result in a quicker reaction.

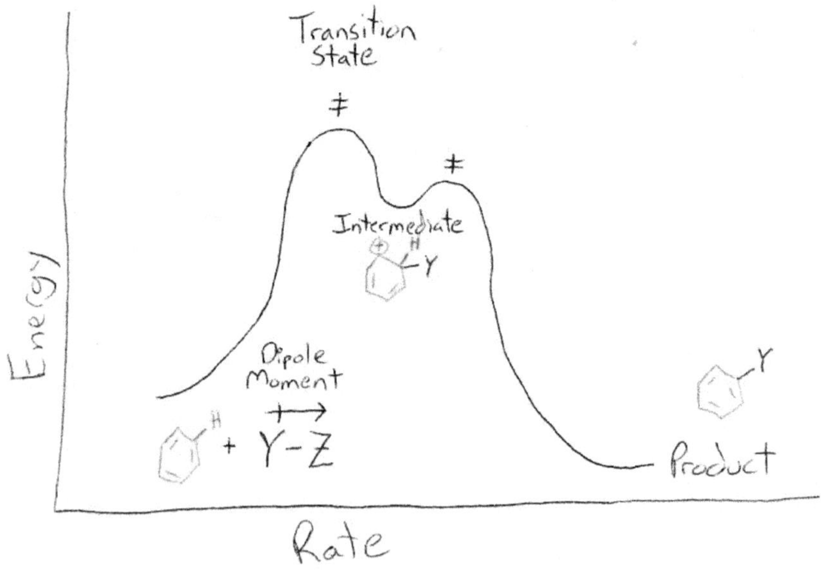


Figure 4: Energy Diagram

All of which is the second time I've reviewed this shit. So no complaining about it. (You can complain if you want to, but I'm just going to pretend not to hear you.) In any event, the 'event' is the RDS. The smaller the RDS, the quicker the reaction and subsequently the more selective the reaction. Or in simpler terms, look at the next figure.

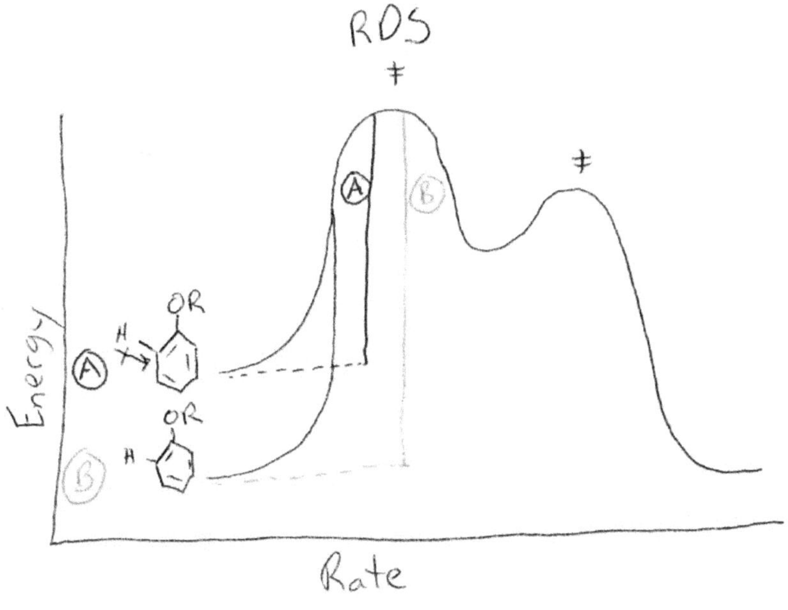


Figure 15: Magnetically Enhanced Synthesis

For simplicity sake, this Energy Diagram is not correct. By that I mean, I have no clue of the absolute variance in the energy between a **proton with a dipole moment** and <u>one without</u>. But, if know that the energy of the Rate Determining Step is determined by the amount of negative charge on the aromatic ring, then a dipole will create more negative charge on the aromatic ring, which will attack Y-Z. Or in more complex terms, any increased negativity at a carbon atom, e.g. dipole moment, will lower the energy of the Rate Determining Step. Or in simpler terms, the line for 'A', blue, is shorter than the line for 'B', red, in Figure 15. All of which means, MORE of molecule 'A' will react, researchers will have to spend less money purifying the ortho/para mixture, and scientists can make cheaper drugs with less toxic waste. And just in case you're still confused, here is the mechanism of the above reaction.

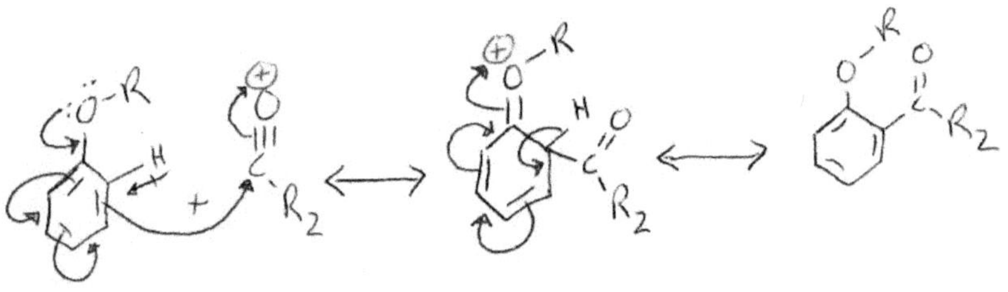

Figure 16: The Mechanism

For those of you who are comfortable with all the arrows, dipole moments, and resonance structures, you might be wondering the following: Does the dipole moment in the intermediate lower the activation energy in the second step of the reaction? The short answer is: Maybe? I mean, the second step of the reaction is the quick step, but the frequency of the tetrahedral proton will be different from the $sp^3$ hybridized aromatic proton. But, if researchers are able to identify the proper radio frequency of the tetrahedral proton and constantly irradiate it to create a dipole moment, then the second mechanistic step will also have a lower energy of activation. Or in simpler terms, making the intermediate tetrahedral proton go ass-UP with radio innervation will weaken the chemical bond and make the second hump smaller in Figure 15.

In conclusion, there are thousands of reactions that are ALL dependent on molecular electronics, which dictates molecular structure and bond strength. Granted, the simple reversal of one proton isn't going to make a perfect reaction, but the cumulative effect of electronic modulation could improve very sensitive reactions. And the improvement of any reaction has real world bling repercussions, whether or not you are willing to imagine it.

# Cliff Notes

1. We are human and we can't see everything. Therefore, we cannot see the whole universe.
2. The universe is made of energy. Therefore, the conservation of energy dictates that the universe consists of intercalated battery systems.
3. Energy degrades and ALL energy fragments are NOT the same. Therefore, the universe is a collection of different energy fragments or branches.
4. Since each universal branch is different, each universal branch contains a unique amount of positivity or negativity.
5. Protons are really large in this branch of the universe. Therefore, our branch of the universe is overwhelmingly negative.
6. When a negative electron collides with a negative thermal energetic quanta, after being expanded by traversing the positive atomic nucleus, a positive component of the electron is destroyed. As a result of this, the electron maintains stability by emitting a negative energetic quanta in the form of a magnetic energetic quanta, photon, and/or thermal energetic quanta.
7. Super conducting magnets create negative magnetic energetic quanta environment via the directional movement and subsequent distortion of atomic orbitals that are perpendicular to the movement of the electrons surging through the super conducting magnet.
8. Protons release negative magnetic energetic quanta to compensate for the destruction of part of the proton's positivity.
9. As a result of atomic orbitals corralling large amounts of negative plasma, protons begin to degrade when they are attacked by this negative plasma, which results in the proton releasing negative magnetic energetic quanta and forming a positive ass.

10. When protons begin releasing negative magnetic energetic quanta, columbic repulsion causes the protons to rotate.
11. As a result of rotation, more positive regions of the proton are exposed to negative plasma in the atomic orbital, which causes a different side of the proton to undergo degradation and subsequent release negative magnetic energetic quanta.
12. Much of chemistry is based upon simple columbic interactions of positive and negative charges. Therefore, when a proton is forced to release its negative magnetic energetic quanta towards an atomic orbital, the proton is showing more 'positive ass'. When a proton shows more 'positive ass' to external molecules, the proton becomes more acidic because there is less columbic repulsion between the proton and external molecules. Subsequently, the proton chemical bond gets weaker when the proton is forced to continually release its negative magnetic energetic quanta towards the negative atomic orbital.
13. When a proton is forced to release its negative magnetic energetic quanta towards an atomic orbital, the proton is forcing more negativity plasma towards the atomic nuclei, which is notated in organic chemistry as a dipole moment.
14. When a dipole moment is directed towards a carbon atom, there is an increase in electron density around the carbon atom.
15. Increased electron density about specific carbon atoms enhances the selectivity of electrophilic aromatic substitutions reactions.
16. Modulation of proton based dipole moments and the amount of positive proton ass shown to NMR enhanced purification matrix, will change the way a molecule interacts with the purification matrix, which will modify the molecules retention time.
17. Given the advent of a chiral NMR machine, which uses a chiral magnetic environment as a discriminating factor, enantiomers can be resolved via a similar methodology, which will be discussed in my next book: *Imperfect Perfection*.

# VI. Response Ability

In a much as the NRA will hate this, someday lasers will replace guns. Granted, the ethics of lasers will be even more complicated than traditional explosive based killing devices because our fore fathers didn't include them in the Constitution, but that is beside the point. The point is, I enjoy thinking about lasers and ethics, so you know: Me book-um another collection of ideas disguised by lots of words with inappropriate grammar and punctuation!?;-) So please refrain from picking your nose, neglecting your children, or farting in public...unless you really really have to.

To the general public, lasers are for correcting vision so they look away from gun violence. But, in the future, people will be confused as to when to look away because lasers will be correcting their vision and killing vast crops of minds that could be used to make the world a better place. All of which, makes me very very excited. (I'm excited for people seeing the world the way it is and NOT genocide.) What will happen when people actually see what is happening in the world around them? Maybe someone will discover a way to numb the consciousness of disagreeable people with lasers without shooting holes in the heads? Or, maybe toe fungus will be declared as a national threat. (Was that a metaphor?) In any event, somebody has to start somewhere if we want society do devolve...*I mean*, evolve. So let's begin.

Light Amplification Stimulated Electronic Release And Decay (LASERAD): Is a process by which decayed-light-energy inundates a matrix of elements to stimulate electron decay and the subsequent release of intense directional light. As you can tell, it is NOT an "Amplification" in the literal sense. By that I mean, more energy is poured into the system then is released. Or in simpler terms, it is like shooting a cow to get the milk. (Not really, but I thought the imagery was funny.) Or in the simplest terms, it is like paying someone fifty bucks to give you three dollars and thirty-three cents in three stacks of a hundred and eleven pennies. Unfortunately, as with most things in science and popular culture, the word LASER is too catchy to be changed. Therefore, I will not get worked-up about science's inability to evolve.

You might not know this, but a little of the Holy Ghost has always been loaded into bullets, which has kept the Holy Ghost really busy. But, God gave the Holy Ghost a promotion without any pay raise. (The Holy Ghost was pissed until he saw the benefits package.) In any event, the Holy Ghost is now squeezed into lasers, which can improve people's eyesight and their ability to ignore the violence in other people's neighborhoods...I think.

So there you have it ladies and gentlemen: An Introduction. I briefly talked about the role of the Holy Ghost in ethical degradation of matter as it pertains to the enforcement of energy responsibility. (FYI, I was totally NOT talking about that?) If this is something that sparks your interest, please continue reading. And if this is something that does NOT spark your interest, keep reading because the Muslims want to kill you with these futuristic laser guns. (Just kidding, I think most Muslims just want to have a family and comfortable life...Unlike the Christians that want to make themselves sick with the food they eat.)

# Chapter 2

Before I get into the exciting world of lasers and poking holes in other people's heads, I NEED to talk about something that has been bothering scientists for quite some time: Gravity redirecting light. Now for the average Joe on the street, this is NOT very bothersome because: Hey, forget about it! No seriously, what were we talking about? Were we talking about bending light, concentrating light, reflecting light, refracting light, or redirecting light? Whatever the case may be or maybe or whichever word collection is correct, I can hear the angry scientists grinding their teeth. Or maybe I'm the one grinding my teeth? Whatever the case is, some people find this question bothersome: How can I start a new chapter without a fucking introduction? Wait, wrong group of people. Some other people find this question bothersome: How are **negative** photons spurred to move <u>towards</u> a massively **negative** star? The religious answer is: The Holy Ghost spurs them. (FYI, 'Holy Ghost in spurs' is my safety phrase.) The simple answer is: Let me take a moment and pick my nose.

Honestly, I didn't have to pick my nose, but I needed a little time to think about this Quanta Quagmire. As a result of having no clue where this jumbled answer may go, please keep in mind that I ALREADY deleted the most illogical answers...hopefully.

1. It just does, so shut your trap.
2. Magnetic fields have been known to slightly redirect light, which is why stars twinkle. (Unless...every star in the universe releases its energy in a perfect five point star format with reference to Earth, which means WE are still the center of the Universe...Booyah!)
3. Photons are afraid of death so they always try and snuggle up to fuzzy stars in an attempt to conserve energy.
4. The Holy Ghost takes a second and calculate the precise angle of deviation as it pertains to photon density, photonic charge, a star's columbic diffusion rate, a star's specific columbic nature, and a star's ability to get a boner on the first date, then spurs the photon in the appropriate direction.
5. **As a result of our star diffusing <u>negative</u> energy for eons, stars can evolve into a positive mass, which can bend negative light. It is important to note that our star's Highly Ordered Energy was depleted a long time ago to yield a**

> **semi-stable mass of positive protons, negative electrons, and negative plasma, which has been slowly degrading and diffusing. (Ironically, this negative diffusion theory gives a method to: Star densification via the fusion of heavier elements in order to lose positive positron charges; Star explosions that occur when stars are unable to lose any more positivity and have already lost too much negativity.)**

As you can see, there are some very **simple** ways to describe HOW and WHY negative photons might SLIGHTLY move towards a negatively diffusive, but highly positive star.

Having said that though, I've got a major problem with this **ONE** observation of HOW stars bend light and HOW this supposedly pertains to gravity: It does not ACCOUNT for **ALL** the photons! (No seriously, does it ACCOUNT for **ALL** the photons?) I mean, the percent error of this calculation is just UNCOMFORTABLE. First, you're staring at the brightest thing in our solar system, which happens to be sending out a crap ton of photons. Next, scientists are calculating gravity based on the angle of deviation of just a few photons of light from a star that is in a galaxy far, far away. (Actually, the star in this probably in this galaxy, but it is still far, far away.) And on top of all that, how would we know if the distant star's photons are being refracted and/or mushed into OUR star's negative light spray? Seriously, how would you differentiate between two different light sources that have ABSOLUTELY the same angle of incidence? Granted, you could check the percent and type of photons, but let's be honest here: You could take the most obscurely colored star in the universe and blend it into our star's light and you would NEVER notice a difference in our star's light because we are really close to our star, our star is putting out a crap ton of light, and our star's light diffusion area is insanely large. Quite simple, it would be like looking for a needle in a haystack if you split a needle into single-atoms and sprayed the atoms all over the haystack. Or in more complex terms, it just does, so shut your trap. In any event, I now feel comfortable enough to continue my conversation about lasers...That is, if you're still in the mood to learn how to poke holes in confrontational craniums.

You remember how I said that a laser was like paying someone fifty bucks to have that person give you three hundred and thirty-three cents in three stacks of a hundred and eleven pennies? Well, I said it in the first chapter somewhere and it isn't quite true. I mean, it is true-ish, but the wibby-wobbly decaying universe causes the statement to be INACCURATE. By that I mean, photons don't squirt out of a laser in perfect little stacks. Or in simpler terms, things are about to get CRAZY complicated.

Do you remember the first book in this collection of books? (You can look, just don't tell anyone.) Well, in the first book, I made this whole hullabaloo about HOW and WHY protons are massive in **THIS** branch of the universe. If you don't remember, protons are massive because all the negative energetic quanta in this branch of the universe makes the positive energetic quanta **expand**. Any who, this is HOW and WHY electrons are able to degrade. You see, when an electron passes through that atomic nucleus is search of that wonderful positive charge, the electron is transported to another dimension where the electron is surrounded by copious amounts of positive energy, which makes the electron **expand**. (Granted, the electron doesn't really get transported to another dimension, but the sentence was missing 'Je ne sais quoi' and everyone seems to be magical thinkers on this planet. So I thought: 'What the hell? I'm going to toss some really crazy shit in this mofo!') As a result of the electron NOT getting transported to another dimension, it has to spend time within a region of **THIS** galaxy that is really <u>**positive**</u>, which makes the electron **expand**. Unfortunately, since the electron has a shitload of momentum and is unable to find the positive charges in the atomic nucleus, the electron flies right though the atomic nucleus. Upon exiting the atomic nucleus, the electron begins to shrink back to its normal size because of all of the negativity in this branch of the universe. BUT, before the electron can return to its normal size, it CRASHES into a cloud of negative thermal energetic quanta, which makes the electron degrade and squirt out a photon. (Remember, the previous logic of electron degradation is as follows: Electrons have some positivity, external negative thermal energetic quanta destroys some of the electron's positivity, and the electron has to squirt out a negative photon to compensate for this loss of positivity.) All of which, complicates laser technology with the following factors:

1) Everything that pertains to electrons: The element they are attached to; The oxidation state of the element; The arrangement and distance between elements; The electron's atomic orbital; How "OLD" the electron is.
2) Everything that pertains to plasma: The density of the plasma; The type of plasma; The negativity of the plasma.
3) Reflection and/or refraction behavior of the laser matrix.

As a result of all these interrelated factors, laser technology has stagnated. So what can be done to improve laser technology such that we ALL have the God given right to walk around with these killing devices in a civil society? Well, since battery technology sucks and nobody wants to strap a generator to their backs, like in Ghost Busters, then the only way to improve lasers is by making them more efficient. (If you actually make it to the last chapter in this book, I will be extremely happy to share my design for handheld lasers, which is totally cool…at least to me.)

The first logical method to making lasers more efficient is by using diodes. You know, that CRAZY energy efficient technology that you can import in your home to decrease world gun violence? Well, the less energy required to stimulate electron degradation, the less energy a laser will require. All of which means, smaller batteries and/or more laser blasts to burn holes in the people you don't like.

In conclusion, EVERY factor of a laser plays a part in the energy efficiency of a laser. Or in simpler terms, we need to outsource our photon construction to third world countries so we don't have to pay the Holy Ghost to protect our morals…OR…We can just understand the theory behind lasers and construct more efficient lasers.

# Chapter 3

You remember how I said that a laser is like paying someone fifty bucks to give you three hundred and thirty-three cents in three stacks of a hundred and eleven pennies? Well, I said it in the first and second chapter and it isn't quite true. I mean, you have to remember photons are negative energetic quanta that are zigging & zagging as a result of innervating with the external negative magnetic energetic quanta in Earth's magnetosphere. Actually, a more accurate description would be: Lasers are like paying someone fifty bucks to give you three Jenga piles of a hundred and eleven pennies. Unfortunately, this statement depends on many factors.

For example, you remember how light degrades into negative thermal energetic quanta in the laser matrix and how ELEMENTS expand when they get hot? Well, that means that each Jenga pile of a hundred and eleven pennies is NOT shaped like a straight missile silo. Instead, each Jenga pile of a hundred and eleven pennies has CURVES, which is dependent on how the Jenga pile is stacked, the external energetic environment, how the photons in the Jenga pile have to refract/reflect around the elements in the laser, and how the Jenga pile of negative photons are repulsed by the other negatively charged photons that are trying to wiggle through Earth's negative magnetic energetic quanta. All of which, gets too confusing. So let's return to my previous example.

You remember how I said that a laser is like paying someone fifty bucks to give you three hundred and thirty-three cents in three stacks of a hundred and eleven pennies? Well, I said it in the first and second chapter and it isn't quite true. I mean, a more accurate description is as follows: A laser is like paying someone fifty bucks to give you three hundred and thirty-three cents in three Jenga stacks of a hundred and eleven pennies where each penny is balanced on its edge in one of 360 possible orientations. All of which, is a great mental description, but physically impossible arrange on Earth. So for the remainder of this chapter, you need to imagine my examples are in space.

Unfortunately, the moment my crazy Jenga example is transported to an environment with a less intense magnetic energetic quanta environment, the example must be revised. Therefore, a laser is like paying someone five hundred dollars to give you nine hundred and ninety dimes in three Jenga stacks of three hundred and thirty **dimes**, where each **dime** is balanced on its edge in one of 360 possible orientations.

The reason for this drastic change in VOLUME is because the photonic volume is directly proportional to the negative magnetic energetic quanta environment. Or in simpler terms, photons on Earth are the size of pennies, photons in space are the size of dimes, and photons in space with NO magnetic energetic quanta are the size of photons. (All of which, gives a great method for the propagation of light though space because regions that are void of negative magnetic energetic quanta will CONCENTRATE negative photons. Or in simpler terms, the conservation of energy dictates that negative photons will concentrate as they move though a region of space that is void of energy, which is the reason why we can see stars that are so far away?)

Hopefully, you see were I'm going with this. If not, then let me take a moment to clarify something: You remember that crazy Jenga dime example I used to describe lasers in space, well it is not quite true. I mean, a laser is like paying someone five hundred dollars to give you three Jenga stacks of three hundred and thirty dimes where each dime is balanced on its edge in one of 360 possible orientations then **having someone take away every dime that is**

**not facing exactly 90 degrees**. As for the reason why someone has to take away all the dimes that are not facing exactly 90 degrees, well that has something to do with polarization. Or in more complex terms, the only way scientists have been able to concentrate photons, because photons bob-n-weave based upon adjacent negative photons and external negative magnetic energetic quanta, is to ONLY allow PARALLEL photon to escape the laser matrix. Or in simpler terms, parallel pancakes stack better. Granted, this is COMPLETELY inefficient, but scientists lack the ability to cause directionally specific electron degradation.

So let me recap, just in case you got lost somewhere:

1. Photons are negative and continually bob-n-weave to avoid other negative energetic quanta.
2. Photons are degraded from electrons, which means **current** photonic density is limited by atomic radii, other matrix factors, and polarization milieu.
3. Negative magnetic energetic quanta can reflect, refract, and/or direct negative photons.
4. The absence of negative magnetic energetic quanta causes photons to concentrate in an attempt to conserve energy, which means the theoretical density of photons are related to their columbic charge.

5. Atomic orbitals are designed to keep SMALL negative entities away from the atomic nucleus as it shuffles its positrons.

All of which, when aligned linearly gives a feasible method to create lasers with such density that you could blast an atomic nuclei with millions of negative photons, which will destroy the positrons and cause atomic fission.

So there ya'll have it: A viable method to creating denser and stronger lasers, which can cause fission. Unfortunately, this technology will be used to punch holes in people and not the advancement of science and humanity. (Dude, just imagine if you could turn a person's head into an atomic bomb...NERDS!) All of which, tears me apart. I had to weight the POSSIBILITY of humans using stronger lasers to turn rocks into fission propulsion systems, which would allow humans to travel the universe, versus the inevitable slaughter of billions of people. (I guess you now know how Nobel felt when he gave TNT to the world.) . But, don't worry. You, as a species, are devoid from any ethical conundrums because you didn't bring these ideas to light...I think.

In conclusion, the vibrational volume of photons are dependent on their directional innervation with other types of energy, which means it is possible to focus photons into denser and stronger laser beams by: 1) Using energy efficient diodes; 2) Using denser elements to emit photons; 3) Using negative magnetic energetic quanta to decrease photon vibrational volumes and to push them closer together; 4) Remove the Holy Ghost from the energy efficiency equation.

# Chapter 4

My human hole puncher (HHP) design is really quite simple. First, it contains magnetic focusing to condense all the negative photons instead of using wasteful polarization medium. Next, it doesn't contain a battery...at least not in the traditional sense. And finally, it has a Gatling gun style collection of crystals, which emits the photons. Hopefully, everything is straight forward except for the lack of battery...in the traditional sense. Therefore, let me take a moment and explain.

You know how negative energy is diffusing from the Earth's core as a result of extreme density? Well, if you create a contained explosion that modulates pressure and temperature, like in the Earth's core, then this device will be able to emit a plethora of negative electrons. Or in simpler terms, it is like a bullet except minus the metal propulsion. Or in the simplest terms, it is like a bullet that has been welded to the casing. In any event, there are numerous aspects to this futuristic weapon that will raise any warlord's eyebrow.

For starters, you don't have to rely on any unstable country to mine projectile bullet material. Or in simpler terms, you won't have to mine all that good metal and then have some dirty third world scum dirty that metal up with their tainted blood. Or in the simplest terms, a cartridge will rotate explosive battery-ullets, which will explode, release enough electrons to power the laser matrix and magnetic focusing system, and then slip back into the cartridge to be recycled. And the best part of all is: Your hatred won't be spewing copious amounts of toxic burnt gun power all over the world.

In any event, an explosion within the sealed Battery-ullet causes an increase in pressure and temperature, which causes atomic orbital re-orbitization and the release of electrons. And as the Battery-ullet cools down in the recycle magazine, the Battery-ullet will suck in some electrons from its environment, which will also provide additional power for your Human Hole Puncher and/or laser armor. So, this weapon is hippie friendly, except when it is used on hippies.

In conclusion, photonic volumes are based upon the external energetic environment. But when you use magnetic focusing, you can *decrease* the volume of photons, **increase** the density of the photons, and <u>decrease</u> the diffusive effect of the external energetic environment. And finally, Battery-ullets will enable any warlord to take out his or her vengeance upon any nation, regardless of their natural resources, because Battery-ullets are environmentally friendly and can be recycled.

# Cliff Notes

1. Electrons expand while traversing the positive nucleus and then crash into negative thermal energetic quanta, which destroys some of the electrons positivity. As a result, the electron maintains stability by emitting a negative photon.
2. As a result of photons being released by electron decay, photonic density is determined by density of atoms in the laser matrix, the polarization milieu, and photonic volumes.
3. Photonic volume is determined by the charge of the photon and its innervation with the external energetic environment.
4. When there is NO negative environment, photons concentrate to conserve energy based upon their negative charge repulsion.
5. When there is a strong NEGATIVE circular environment of negative magnetic energetic quanta, the negative magnetic energetic quanta will decrease the photonic volume and force the negative photons to move closer together.
6. Denser photons will contain the momentum and charge to punch through atomic orbitals and destroy positrons, which will lead to fission.
7. The end of this book may facilitate the end of civilization UNLESS more people stand up to promote the pillars of civilization: Respect, Tolerance, and/or the Conservation of Energy.

www.ingramcontent.com/pod-product-compliance
Lightning Source LLC
Chambersburg PA
CBHW051649170526
45167CB00001B/390

* 9 7 8 1 5 0 5 5 0 1 0 6 3 *